THE INVESTIGATIONS BOOK

A resource book for teachers of mathematics

JOHN HOLDING

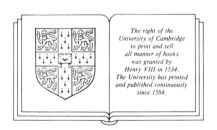

The right of the
University of Cambridge
to print and sell
all manner of books
was granted by
Henry VIII in 1534.
The University has printed
and published continuously
since 1584.

CAMBRIDGE UNIVERSITY PRESS

CAMBRIDGE

NEW YORK   PORT CHESTER

MELBOURNE   SYDNEY

Published by the Press Syndicate of the University of Cambridge
The Pitt Building, Trumpington Street, Cambridge CB2 1RP
40 West 20th Street, New York, NY 10011–4211, USA
10 Stamford Road, Oakleigh, Melbourne 3166, Australia

First published 1991

Printed in Great Britain at the University Press, Cambridge

British Library cataloguing in publication data
Holding, John 1928–
Mathematical investigations.
1. Schools. Curriculum. Mathematics. Teaching
I. Title
510.71

ISBN 0 521 40944 6 paperback

VN

CONTENTS

INTRODUCTION

In order to advance, whether as a learner or an investigator, it is necessary to jump to conclusions, and of course it is equally necessary to check the conclusions. Charles Godfrey, 1911.

The teaching of mathematics should acquaint the students (pupils) with all aspects of mathematical activity . . . especially it should give opportunity . . . for independent creative work. George Polya, 1964.

Pupils should use number, algebra and measures, shape and space, and handle data, in practical tasks, in real life problems, and to investigate within mathematics itself. National Curriculum, 1989.

Godfrey and Polya are two giants of twentieth-century mathematical education who more than any others have profoundly influenced my approach to mathematics teaching. They are both fully committed to the belief that problem solving (some say 'resolving') and problem posing are at the heart of mathematical learning. Clearly their counsel has now achieved national recognition.

This book attempts to provide substantial evidence on how National Curriculum Attainment Targets 1 and 9 (quoted together above, but see note at the end of this introduction) are realizable in practice; it principally consists of commentaries on investigations which I have used over the last twenty-five years. The ideas outlined in the commentaries are, for the most part, those developed or suggested by pupils at primary and secondary levels, by students and by teachers (both by themselves and as a result of using the investigations with their own pupils). All their contributions are acknowledged with gratitude. They enjoyed creating mathematics, original at least to themselves, and achieved a sense of fulfilment in so doing.

The intention is to show what can be achieved from specific starting points: though some problems are solved, others are posed as a consequence. It is *not* to suggest that there is only one way of developing an investigation; nothing could be further from the truth. Only those approaches which have been successfully developed are discussed here but there can be no doubt that other approaches are quite possible and very likely if investigators are encouraged to develop their own lines of thinking. It is intended that teachers should use the commentaries as a benchmark to compare with the results achieved independently with their own pupils.

All of the investigations have a statement of problem, that is the starting point; these are designed both to be readily understandable and to contain an immediate element of challenge. The first group of 18 investigations then have detailed commentaries with sections of general comment, possible approaches and extensions. Full detail on development has been possible as a result of wider experience of using these particular investigations. The second group of 22 have been less widely used; they just have sections of comment and suggestions which briefly sketch some ways in which the investigation has been developed and other relevant points. Within each group the investigations are in a very approximate order of increasing necessary background knowledge.

A few of the investigations will be well known and have become classical in the established literature on the subject over the last ten years. They have a type of eternal quality which appeals to all on first acquaintance.

Computing, whether in the form of commercial software or simple pupil-written programs, can frequently enhance the investigation process to a marked extent.

Preceding the commentaries are five short chapters concerned with sources of investigations, the nature of mathematical thinking, and with practical problems of implementing investigational work with pupils and students. Chapter III, on mathematical thinking, contains many references to the National Curriculum, in particular to Attainment Targets 1, 5, 9 and 10.

All the Attainment Targets (ATs) are examined in some detail in Chapter V, which includes a table of cross-references to all of the investigations. Further, within each investigation commentary, a short section has appropriate references to the relevant ATs; these are chiefly to ATs other than 1 and 9 and identify the minimum level of knowledge and concepts necessary for and likely to be enhanced by the exploration of the stated problem.

All pupils and students from this level upwards should be able to tackle the problem to their particular level of attainment in ATs 1 and 9. Only when this latter level is higher than the previously identified minimum is there a direct reference to ATs 1 and 9. The full development described in the commentaries covers aspects of pupils' and students' work from the minimum level upwards.

There are also suggestions for basic materials likely to be needed.

Note

In May 1991 the Secretary of State for Education and Science published proposals for revising the presentation of the National Curriculum with the objective of simplifying the structure and making assessment procedures more manageable. The Attainment Targets have been grouped together into five NATs as follows: ATs 1, 9→NAT 1; ATs 2, 3, 4, 8 (part)→NAT 2; ATs 5, 6, 7→NAT 3; ATs 10, 11, 8 (part)→NAT 4; ATs 12, 13, 14→NAT 5.

Within each target there has been some minor editing of the statements of attainment, now called programmes of study, with the new statements of attainment being a summary of the programmes of study. The material at each level within each target has not been altered.

The majority of references in this book to the National Curriculum are to the old statements of attainment within specific ATs. The references can be easily related to the new proposed structure using the above table.

I WHY INVESTIGATE?

The principal activities of investigation, observation, collection of data, search for pattern, conjecture, verification, insight, analogy and proof are not new to mathematical learning. What would appear to be new is the realization that thinking skills such as these can only be acquired and encouraged by the involvement of pupils and students in investigative work as an end in itself. The problem for the future, therefore, is how to integrate and interrelate this type of activity with the established mathematical curriculum.

HISTORICAL REVIEW

A brief review of prominent developments throughout the century towards the present position seems appropriate. The pattern of mathematics teaching which was current during at least the first half of the present century is that of imparting as much as possible of a given body of knowledge and manipulative skills, tailored to the ability of the pupils and in the most educationally efficient manner possible.

Within this context many teachers of mathematics had realized that any topic in the mathematical curriculum at all levels can be enlivened and enriched by an investigative approach in the first instance. Examination of some of the educational writings of Charles Godfrey, of the Mathematical Association reports on the teaching of various topics, of articles in *Mathematics Teaching*, the journal of the Association of Teachers of Mathematics, and of textual material such as that of the School Mathematics Project or Midland Mathematical Experiment produced in the 1960s reveals this clearly. In 1978 the South Nottinghamshire Project even went so far as to devise a two-year investigation-based curriculum scheme for pupils aged 11–13 which also claimed to cover the content of normal syllabuses.

THE PROBLEM METHOD

An early example of this approach to the construction of triangles uses the following problem.

On my(!) estate there are three trees: a beech, a willow and an oak. I wish to draw a plan showing them. A lake wider than the length of my measure separates the beech and the willow. What measurements must I make in order to draw the plan? (All tricks with boats and string are barred.)[1]

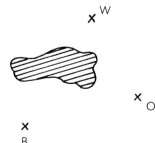

There is an air of artificiality about this problem which may stem from it being concocted to serve a particular end, that of the necessity for measuring the 'included' angle (a non-included angle giving rise to possible ambiguity). Nevertheless the desire is genuinely to relate a mathematical result to a practical problem and to lead the pupil to a chosen knowledge goal via the processes of enquiry and experiment.

Underlying the transmission of a particular item of mathematical knowledge the authors (teachers) advocating this approach to the teaching of geometry had another laudable aim for their pupils, that of being able to understand the significance of particular methods and results and to appreciate their elegance and simplicity. Such teachers were already acting on the principle of paragraph 250 of the Cockcroft Report long before that was written.

> The idea of investigation is fundamental to both the study of mathematics itself and also to an understanding of the ways in which mathematics can be used to extend knowledge and to solve problems in many fields.[2]

The paragraph goes on to encourage the use generally of the investigative outlook in teaching, ranging from short episodes to protracted studies. This aspect of mathematics teaching has an analogy with the parallel subject to mathematics in the curriculum, that of the study of language, which has amongst its prior aims that of introducing pupils to the pleasure to be found in reading literature and in creative writing.

However, although these investigative approaches were employed, most schemes of evaluation and assessment at all levels, both internal and external, were by their nature unable to take much, if any, account of them; indeed some teachers argued that to examine the creative and problem-solving aspects of mathematics learning is impossible. From the 1950s onwards however some attempts have been made to widen the scope of assessment; the CSE allowed for innovation in assessment and there are a few isolated examples of non-CSE examinations including an element of continuous assessment, normally in the form of projects or investigations. At the higher education level, coursework elements are now frequent in B.Ed and PGCE courses, most of which are taught by teachers with previous and ongoing involvement at primary and secondary level. From 1988 these developments have been regularized on a national basis at secondary level with the criteria for the GCSE requiring investigational work to be assessed.

THE NATIONAL CURRICULUM

With the coming of the National Curriculum[3] (see Chapter V) investigational skills have become a statutory requirement in the mathematical curriculum for ages 5 to 16; Attainment Targets 1 and 9 are about using and applying mathematics and specifically list such thinking skills as 'make and test predictions'. Further references can be found in Attainment Target 5 which is concerned with recognizing and using patterns, and making generalizations.

Assessment arrangements for the National Curriculum are however as yet incomplete; at age 16, the expectation is that they will largely follow GCSE lines already established.

THE PRESENT SCENE

The second half of the twentieth century has thus seen a gradual growth in the realization by teachers that not only is it important and educationally desirable that their pupils should understand, appreciate and be able to apply the content of the syllabus but that the acquisition of specific thinking and experimental skills should be encouraged by problem-solving activities both independent of and related to the content syllabus. In the words of George Polya not only have we to use our heads to learn the subject, but use the subject to learn to use our heads.

Problem-solving is bound to be related to knowledge in so far as any problem-solving activity requires prior knowledge of some mathematics; the question of its relation to knowledge to be acquired in future is much less clear. The knowledge versus thinking-skills debate is going on in several other subjects of the curriculum as well; in mathematics the ideal curriculum would seem to be one in which the two strands are fully interwoven, though the objectives at any particular stage *must* be clear.

PROGRESSION IN LEARNING

We have tended to think that knowledge acquisition can be strictly organized on a time basis, that a specific step forward can be achieved in a specified time, typically that of a lesson or seminar in the timetable. This view is quite false; much research has shown that the mapping

time spent in study → knowledge acquired

is far from linear. Equally, thinking-skills cannot be acquired on an organized linear time-base; perhaps this seems more obvious at present, though it may be because we lack wide enough experience in framing a thinking-skills/problem-solving orientated curriculum. It would be unfortunate if in due course we fall into the same linear time trap with thinking skills as we do now with knowledge acquisition.

REFERENCES

1 Mathematical Association (1938) *Teaching Geometry in Schools* Bell, p.10.
2 DES (1981) *Mathematics Counts* (Cockcroft Report) HMSO, para 250.
3 DES (1989) *Mathematics in the National Curriculum* HMSO.

II SOURCES OF INVESTIGATIONS

All investigations develop from an initial problem, the starting point. Such a problem has several qualities which make it a problem to a particular person or pupil. To that person it must

- be intelligible,
- have motivation and challenge,
- contain blockage, that is the solution is not immediately apparent,
- imply some discrimination between possible courses of action.

CREATING A STARTING POINT

Framing a starting point, in either written or verbal form, and with the above qualities, needs care. A good technique is to describe initially a particular case of a problem, with a diagram if appropriate, and then pose a question or a sequence of related questions in such a way that a response is possible, either immediately or within a short space of time. This type of technique has been employed in the starting points for the investigations analysed in this book.

For example ISOLATIONS (4) starts:

Here is a chain of the
first four whole numbers:

In this chain 1 and 4 are **isolated** because neither has its normal numbers next to it; 2 and 3 are not isolated.

Can you make a chain of the same four numbers so that they are all isolated?

Because the number of arrangements of 1, 2, 3, 4 is limited this initial question will soon be resolved, even if some systematic method is not employed. The problem is now firmly established in the mind of the investigator who should then be invited either to find more examples which fit the given requirements or to vary the conditions in some respect.

ISOLATIONS (4) continues:

Experiment with different sets of numbers and different sorts of chains.

This suggestion immediately throws open the problem and a wide variety of developments is possible. It might be advisable to make this experimental stage more specific than the above wording suggests.

With this technique for framing investigations the starting point has given the investigator both an immediate result and an involving activity. To use an analogy the starting point is the device which activates the engine which in turn enables progress along the investigative road.

For some investigations motivation can be enhanced by putting the problem in the context of some actual or supposed real life or exciting situation. Dressing a classical unsolved problem in the format used in CHAIN TRAPS (1) increases interest significantly; POSTAGE (7) is more easily understandable in its real life format than if it had been framed as a pure number problem.

The language employed and the level of mathematical understanding assumed will clearly need to be adapted to match the target user. All the statements of problem in the investigations in this book may need to be and have been on occasions adapted for particular groups of pupils. In their present form they are worded to suit the more-able upper primary pupils (level 5) and upwards. The great majority of teachers who have used them have had no difficulty getting started themselves and have been able to adapt the wording where necessary for their pupils.

WHERE DO SUCH STARTING POINTS COME FROM?

There are a variety of possible sources. In the first instance the journals and publications of the professional associations (Mathematical Association and Association of Teachers of Mathematics) are always a fruitful source of ideas and cross-fertilization between teachers. Articles may contain material, which though not necessarily designed as investigations, can be adapted to create suitable starting points.

The corresponding American journals (from National Council of Teachers of Mathematics and American Mathematical Society) have also been found to be a useful source.

EXPLORATORY APPROACHES TO SPECIFIC MATHEMATICAL CONTENT

It has already been remarked that many teachers and some mathematical texts have employed an exploratory approach to knowledge acquisition. The investigation LIGHTHOUSES (18) is a case in point as it grew out of an exploratory approach to angle properties of a circle; it then developed a momentum of its own which demanded a willingness and confidence on the part of the teacher firstly to encourage the spirit of enquiry and secondly to relax time constraints to some extent, both being essential principles of the investigational outlook.

This is a ready source for investigations in the first instance drawing on familiar areas of the mathematics curriculum; other examples are PIVOT POINTS (37) which developed from considering centres of rotation and SQUARE SPIRAL (39) from simple considerations of limiting processes.

One difficulty associated with this source of investigational material is that successful problem-solving activity requires a good command of the mathematical knowledge involved.

However, the knowledge acquisition dimension of the mathematical curriculum largely operates at the front end of the learner's mathematical experience. With this proviso in mind the mathematical context of this source may need to be simplified to make the investigation readily accessible and workable.

More teachers in the past would have developed this type of experimental approach related to knowledge acquisition had they seen it fitting into existing schemes of assessment.

MATHEMATICAL PUZZLES

These have been around for as long as mathematics, but by the way they are framed they tend to be closed problems rather than investigations. Some remarkable books such as those by Rouse Ball[1] and Beiler[2] abound with possible investigation sources provided licence is taken to vary the conditions of the numerous problems therein. The investigation MAGIC (29) has been developed from the traditional magic square problem by changing the square first to a triangle, then a pentagon and so on, thus discovering possibly original results.

The use of the word 'recreation' in the title of both these books is indicative of an outlook which sees some mathematics as serious and some as comparatively trivial, just for enjoyment. This outlook is not dissimilar to the traditional place of puzzles in the mathematics classroom, an end of term activity when the serious business has been concluded. One possible explanation for this outlook is that seriousness is equated with usefulness, say in physics or engineering; number theory by contrast appears to have few if any real life applications and, consequently, is not serious mathematics.

With the recognition that developing thinking skills is as serious a part of the mathematical curriculum as, say, learning calculus, the traditional puzzles suitably opened up have acquired a new importance as a valuable source of material on which to exercise such skills.

CLASSICAL UNSOLVED PROBLEMS

These should not be neglected simply because they are unsolved. CHAIN TRAPS (1) is based on a fascinating unsolved problem in which a great deal of interest can be found in exploring the patterns that arise. Furthermore, varying the conditions of the problem reveals some solvable problems and analogies with the original unsolved problem.

Prime numbers are traditional sources for speculation, with some known results, for example that every prime number is adjacent to a multiple of 6 (is the converse true?), but also some famous unresolved conjectures such as that of Goldbach that 'every even number greater than 2 can be expressed as the sum of two primes'. Though unproven this conjecture is well worthy of exploration.

For example, we could use the framing technique already outlined and say:

22 can be written as the sum of two prime numbers in three different ways:

$$22 = 3 + 19 = 5 + 17 = 11 + 11$$

Find some more numbers which behave in this way.

Explore patterns in the way even numbers are written as the sums of primes.

Though I have not much evidence on the way this investigation might develop some observations are intriguing. For example, a number as large as 68 can be expressed as the sum of two primes in only two ways ($7 + 61$ and $31 + 37$), while 128 seems to be the largest number with just three ways. Multiples of 30 by contrast always seem to have a high number of sums relative to their size; 60 has 6, 90 has 9, 120 has 12, 150 has 12, 180 has 14 and 210 has 19. Comparatively, even numbers immediately before and following these multiples of 30 have few sums, 58 has 4, 62 has 3, 88 and 92 both have just 4 and so on. There surely must be some reason for these observations.

There is much useful mathematical thinking involved in exploring an unsolved problem in this way. The nature of inductive evidence is well illustrated; as an even number increases in size so does the number of possible representations as the sum of two primes, though highly erratically. It seems most unlikely that a number greater than 68 will have just two sums, let alone a number having none at all; nevertheless it might just happen and Goldbach's conjecture turn out to be false. Who knows?

CLASSICAL INVESTIGATIONS

There are several classical investigations in this book, which have become so because of their immediate universal appeal and accessibility. For example it would be unlikely to find a group of pupils or students not readily fascinated with POCKET (8) particularly if one of the several available software programs such as *Snook* is used. The simple relationship between the dimensions of some tables and the number of rebounds is often conjectured fairly quickly, but there are problem cases to deal with and these provoke much speculation. The investigation is rich in basic mathematical ideas such as ratio and proportion, similarity, symmetry and reflection. Many interesting extensions to the initial problem are possible.

Experience would suggest that because of these qualities this particular investigation, and others in the same category, have almost an inexhaustible capacity. Pupils can explore them on several distinct occasions focusing on a different aspect each time. A teacher can use them again and again with different groups of pupils, frequently discovering some new feature as a result of their pupils' explorations.

REAL-LIFE SITUATIONS

It has already been remarked that framing an investigation in a real-life context can often act as a motivator.

Mathematical exploration of real-life problems has become known as modelling, and it is clear that modelling and investigations share considerable common ground in the curriculum. The difference seems to be one of objective; the objective of investigations is to develop specific thinking skills in mathematics, possibly up to the nature of proof; that of modelling is to apply mathematics to practical problems. It would be too simple to suggest that investigations belong to pure mathematics and modelling to applied mathematics but nevertheless the distinction is useful.

A problem-solving outlook will frequently see possible investigations in real life. However virtually all real-life problems have no clear and easy solution and will need to be simplified before mathematics can be employed to model the problem.

For example, the problem of siting a post box can be explored mathematically, and can be simplified by considering a small village with a number of houses all sited along one side of a road. What criterion should be used? Should it be placed so that the average distance that the villagers have to walk is as small as possible?

A possible solution can be found using this criterion. Is it a satisfactory solution? How can it be modified, if thought to be unsatisfactory for one reason or another? What other criteria should be considered?

A fuller treatment of modelling in the curriculum is due to Burkhardt[3].

REFERENCES

1 Rouse Ball, W. (1939) *Mathematical Recreations and Essays* (11th edn.) Macmillan. (This delightful book was first published in 1892 but has had numerous editions since then.)
2 Beiler, A.H. (1964) *Recreations in the Theory of Number* Dover.
3 Burkhardt, H. (1981) *Real World and Mathematics* Blackie.

III   MATHEMATICAL THINKING

At one extreme of mathematical activity, that of mathematical research, the investigative or scientific method of experiment, observation and hypothesis is the only appropriate method in the first instance. Deductive thinking from axiom to theorem characterizes finished mathematics, though fundamental assumptions can still be questioned. For example, Euclid's analysis of geometry was, for many centuries, generally regarded as finished mathematics, even though doubts were often expressed about the independence, or otherwise, of his fundamental axioms. However, mathematicians in the nineteenth century started very seriously to question some of his assumptions about order, parallels and continuity, and eventually created non-Euclidean geometry.

PROBLEMS, SOLVED AND UNSOLVED

It would be quite wrong to suppose that the only problems which remain unresolved lie at the frontiers of advanced mathematical research; mathematics at all levels abounds with unsolved problems. In this respect the traditional mathematics curriculum has tended to give quite the wrong impression about the true nature of the subject. Furthermore some of the classical unsolved mathematical problems require very little mathematical knowledge to comprehend them. An example is Goldbach's conjecture which has already been discussed in Chapter II. The commentaries on the investigations in this book will reveal a wealth of simple unsolved problems; attempting to resolve these raises numerous other problems. Considering the very well-known chain problem 'if odd, multiply by 3 and add 1; if even divide by 2' gave rise to the investigation CHAIN TRAPS (1) merely by a slight adjustment to the chain condition. Further slight adjustments give rise to a wealth of investigative possibilities as the commentary reveals. Here we have what Eric Wittman[1] calls a problem complex, a neighbourhood of related problems, where one problem spawns many more.

MODES OF MATHEMATICAL THINKING

The National Curriculum divides 14 Attainment Targets (ATs) in mathematics into ten sequential levels. In particular ATs 1 and 9, concerned principally with thinking skills, start at the level of using and talking about materials in an organized fashion, and finish at the level involving some form of proof. The following analysis is more concerned with specific modes of thinking which are not necessarily sequential but which nevertheless reflect many aspects of the ten levels of attainment. Furthermore they examine other aspects of mathematical thinking not explicitly mentioned in those levels.

PATTERN RECOGNITION AND CONSTRUCTION

The ability to recognize and construct organized, as opposed to random, arrangements of objects, such as numbers of two- and three-dimensional shapes, is fundamental in mathematical thinking. Very young pupils, for example, use squares and/or equilateral triangles to create geometrical patterns; such an activity is referred to in AT10, level 2, and is an early form of generalization.

Level 3 of AT5 goes a step further and refers to *explaining* patterns in number; an ability to observe and explain a pattern at this level could be regarded as the first evidence of inductive or even deductive thinking. An example of this could be explaining, in some logical manner, why the even numbers in the standard 100-square form vertical columns whereas multiples of three form a diagonal pattern. If the standard 100-square is then replaced by a snakes and ladder type 100-square in which alternate rows increase in opposite directions, the even numbers appear in a chequerboard arrangement, and pupils can explain why this different arrangement of even numbers occurs. Operating at this level the pupils are beginning the development of their thinking skills.

GENERALIZING

AT5 refers to generalization specifically in the context of algebra while, in fact, generalization occurs throughout *all* areas of mathematics. Polya[2] defines generalizing as moving from considering a set of objects to a larger set containing the given one.

An early form of generalization involves the creation of examples which satisfy some given specifications, an activity which occurs repeatedly in investigative work; the second stage of this form of generalization then involves classifying such examples systematically to ensure that the set is complete. The investigation DELTOMINOES (3) illustrates the two stages; the ability to create figures made up of, say, five equilateral triangles joined edge to edge is the first stage; to then classify them in such a way as to know there are just four distinct possibilities is the second stage.

Further forms of generalization include the ability to extend patterns in number sequences; such sequences frequently arise in the development of an investigation. Some have neatly formulated arithmetic or algebraic rules to define them which can often be explained by reference to the context from which they emerge, for example in the investigations WALLS (5) and CHOP (21); others resolutely defy easy methods of definition in spite of there being a well-organized structure in the way in which they are created; see for example the commentaries on the investigations DELTOMINOES (3), ISOLATIONS (4) and LOOPS (28). At level 3 (AT5) pupils are able to predict the next term of a sequence and, at level 6, to determine possible rules for the generation of a sequence, this being just a more sophisticated form of the same ability. Symbolic form of such rules is seen as level 7.

In geometry too relationships can be generalized. The tessellation of equilateral triangles is generalized into the tessellation of any scalene triangle (which can be convincingly demonstrated by creating a shadow under parallel light projection). Likewise a square tessellation becomes one of parallelograms. A further generalization of this pattern is that any quadrilateral will tessellate, a result which, when first suggested, is often strongly doubted by those unfamiliar with it.

Specializing is the reverse of generalizing, moving from considering a set of objects to a subset. Such special cases are more easily resolved and in turn can throw light on the corresponding more general case. For example, in DELTOMINOES (3) it is easier to enumerate and possibly generalize that subset of shapes formed by equilateral triangles which have a line of symmetry, whereas the number of asymmetric ones proves much more difficult to predict.

ANALOGIES

A frequent strategy employed when investigating is to examine simpler, related, cases of a given problem; these are not strictly special cases but analogous cases. Such analogous cases can be resolved more easily and this in turn builds confidence in tackling the more difficult case, and may even suggest possible lines of attack. Curiously the National Curriculum has no explicit reference to analogy.

Polya states that 'two cases are analogous if they share a common generalization'. Wittman[1] illustrates this point ingeniously by suggesting that we consider the following problem.

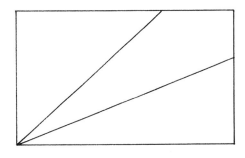

> Divide a rectangle into *three* parts of equal area by two lines through one of the corners.

The problem of dividing the rectangle into *two* equal parts by one line through a corner is a related, or analogous, simpler problem, and in fact is hardly a problem at all! On the other hand, changing 'rectangle' to 'square' in the problem as first stated is an example of specializing. (See the investigation THIRDING A SQUARE (16).)

Another analogous but more complex problem could be posed.

Divide a rectangle into three parts of equal area by *any* two straight lines.

The two lines can be drawn parallel to a pair of opposite sides of the rectangle; this special case has two forms which when combined divide the rectangle into nine equal parts; then using the other analogous problem already considered a solution to the initial problem is revealed as in this diagram.

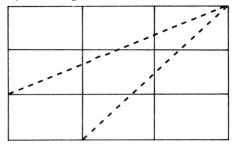

Thus simpler analogous and special cases have been synthesized to solve the more difficult given problem.

A looser definition is that two systems are analogous if there exists a clearly definable relation between their respective corresponding parts; equivalence of fractions is thus an analogy.

Mason *et al.*[3] describe a game between two players.

> Nine cards labelled with the numbers 1, 2, 3, . . ., 9 are laid face upwards on a table and each player takes one card in turn. The winner is the first player who has, amongst the cards in his hand, three which total to 15.
>
> How many different sets of winning cards are there? Develop some winning and defensive strategies.

Of what familiar game does this one remind you? Where is the analogy?

Analogies frequently occur in geometry between two and three dimensions; for example the analogous result in three dimensions to Pythagoras' Theorem, referred to in level 7 of AT10, forms the basis of the investigation QUADRUPLES (34). An alternative analogy to the theorem would be to change the shape described on each side, for example 'the area of the semi-circle on the hypotenuse of a right-angled triangle . . .'.

Polya[4] demonstrates the power of using analogy between dimensions in geometry in his analysis of the 'five-planes problem' which appears in the three-dimensional development of the investigation CHOP (21).

Another geometrical analogy relates the triangle and tetrahedron, namely that each is bounded by the minimum number of bounding elements, lines in two dimensions and planes in three. What is the corresponding one-dimensional analogy? The triangle and tetrahedron have several analogous properties, for example:

$$\text{area of triangle} = \tfrac{1}{2} \text{ base} \times \text{ perpendicular height};$$
$$\text{volume of tetrahedron} = \tfrac{1}{3} \text{ base area} \times \text{perpendicular height}.$$

In the same way as analogous objects have analogous properties, analogous problems often throw light on each other, so the ability to form analogies is clearly one to be fostered. In particular, possible extensions to an investigation are normally some form of analogous problems.

INTUITIVE THINKING

The experience of suddenly 'seeing it' is one which we all, from time to time, have in mathematical thinking. Crossword solvers will be familiar with the experience, particularly in solving anagrams. A sudden insight gives considerable satisfaction and provokes a desire to communicate and, for both of these reasons, we should attempt to provide opportunities for pupils and students to experience such insights.

Poincaré, a French nineteenth-century mathematician, maintained that the rôle of the unconscious in mathematical invention is uncontestable. However, it is the least definable characteristic of thinking processes (perhaps mathematical catastrophe theory could help here) and therefore not surprisingly receives no mention in the National Curriculum.

For sudden insight to occur, firstly it is clearly necessary to fully understand the problem being considered. Secondly it would appear necessary to have confidence with the mathematical knowledge involved in the problem, and thirdly an atmosphere free of obvious time constraint seems essential. Further, a period of rest away from a problem ('take the dog for a walk') often results in new insight, though not necessarily of the sudden variety.

INDUCTIVE THINKING

This is the level of thinking at which **conjectures** or **hypotheses** are formed and tested; generalizing, which has already been discussed, also has the characteristics of forming conjectures. Testing either supports or refutes a conjecture; if refuted then, possibly, the initial conjecture can be modified to take into account the causes of the refutation.

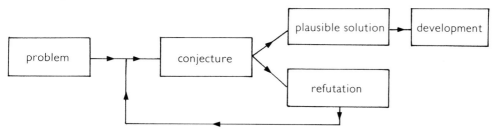

In ATs 1 and 9, level 3 refers to making and testing predictions, level 4 to testing statements and level 5 to making and testing statements; level 6 refers to making and testing generalizations and simple conjectures; a clear progression is envisaged from prediction to statement to generalization and conjecture.

Conjectures figure more strongly in some investigations than others; in DIAGONAL (6) a conjecture, about the number of squares the diagonal passes through, is usually soon suggested, 'length + width − 1', but then found not to work where the dimensions have a common factor. Just one such 'counter example' is enough to destroy the conjecture. Obviously the conjecture can then be restricted by a proviso 'if the dimensions have no common factor then . . .'. Since no examples are forthcoming to refute this revised form it becomes more plausible, though not proved.

The next stage is to attempt to form a conjecture for the cases where the dimensions have a common factor; in due course it should be possible to frame a plausible conjecture to cover all cases. If such a conjecture can be found the innate structure of the problem seems to suggest that it must be more than plausible; what exceptional cases other than the diagonal passing through corners of internal squares could possibly arise? Is the conjecture now proved?

In formal logic terms consider the relationship between two statements A and B. Suppose that B is a *consequence* of A (a conjecture), that is 'if A is true then B is true'.

Inductive reasoning involves examining the truth or otherwise of the consequence of a conjecture; if B is true then A is not proved, it just becomes more credible. The majority of scientific enquiry follows this form of argument; scientific theories are formed as a result of observation and then verified by testing; such theories are never proved in the rigorous mathematical sense.

However a scientific theory can be rejected; if the consequence B turns out to be false then the conjecture A is false; this is **deductive** reasoning, the outcome of which is absolutely clear.

Two other forms of *inductive* reasoning are given below.

- Examining a possible *ground* for a conjecture, on the truth of which the conjecture depends; if such a ground turns out to be false then the conjecture becomes less credible.

- Examining a *conflicting* conjecture to the one under consideration; since both cannot be true, if the conflicting conjecture is false the given conjecture becomes more credible.

Both these types of argument clearly also have deductive forms, namely when a possible ground or a conflicting conjecture are true.

PROOF

Any reference to explaining an observed pattern or result is a form of proof; the word **explain** first appears at level 3, but references to **reasoning** appear only from level 6 upwards.

Basically the proof stage is reached when one is convinced of the truth of some conjecture or generalization, but, how do you convince someone else? Many of the explanations in the commentaries on the investigations are convincing though not in the form of a rigorous proof.

Reasoning requires an ability to follow logical arguments, including those already discussed under inductive thinking. Further the ability to detect flaws in arguments is essential. Mathematics is a useful medium to develop the ability of logical consistency as it does not contain the emotive aspects which can frequently colour argument in some areas of the curriculum and lead to illogical reasoning.

Two important forms of illogicality derive from the statement

if A is true then B is true.

It is often suggested that

if B is true then A is true

and

if A is not true than B is not true

are equivalent statements to the given one; neither would be correct. One need only consider the true statement

if ABCD is a square then ABCD is a quadrilateral

or

all squares are quadrilaterals.

On the other hand

if B is not true then A is not true

is an equivalent statement;

if ABCD is not a quadrilateral then ABCD is not a square

is clearly correct.

A powerful form of proof in mathematics is that of **proof by contradiction**, mentioned in the National Curriculum at level 10; the inductive argument using the examining of two conflicting conjectures becomes deductive if the two conjectures are the only possible ones.

A classic example of this is the proof of the irrationality of the square root of 2, by assuming that it *is* rational and showing that this leads to an impossibility or contradiction.

At the beginning of this chapter, axioms are mentioned; in logical argument truth has to begin somewhere and as such cannot be proved. We teach our pupils to unconsciously, though sensibly, accept the real number system as axiomatic; explanations of the many conjectures and generalizations which arise about numbers (see for example the investigation STAIRCASE NUMBERS (2)) unconsciously assume these axioms. In geometry, it is not so obvious what should be axiomatic, and perhaps the Euclidean model has too much dominance.

REFERENCES

1 Wittman, E. (1975) Matrix strategies in heuristics, *International Journal of Mathematical Education*, Vol. 6/2.
2 Polya, G. (1954) *Induction and Analogy in Mathematics* Princeton, ch. 2.
3 Mason, J. *et al.* (1982) *Thinking Mathematically* Addison Wesley, p. 88.
4 Polya (*op. cit.*) ch. 3.

IV   IMPLEMENTING INVESTIGATIONS WITH PUPILS AND STUDENTS

It has already been remarked in Chapter II, Sources of Investigations, that careful framing of a written or verbal starting point for an investigation is crucial to its successful use. This is the first stage in implementing investigations in the curriculum, but much more needs to be considered about how to get going, teaching strategies, the rôle of the teacher, classroom layout, time constraints, the relationship with other aspects of the mathematics curriculum, and assessment.

GETTING GOING

If investigations are a new experience for the pupils or the teacher or both then, inevitably, there are some initial hurdles to overcome. Teaching and learning styles are very different from those appropriate to a mainly instructional mode of learning. Though the teacher remains the initiator and overall controller, much of the decision-making is transferred to the learner, and pupils' attitudes and expectations will require enlarging in order for them to respond effectively to this new alternative way of working.

Pupils' first experiences of investigative work must be engaging and enjoyable. This is a vital step towards ensuring that they find the new way of working acceptable and have positive attitudes towards it. Using investigations such as DIAGONAL (6) or POCKET (8) on a class basis invariably proves successful with previously uninitiated groups. The starting problem is rapidly understood and a class can quickly provide a variety of data which could be collected together in some agreed systematic manner on the blackboard. Observations and speculations about this data then readily follow. The commentaries on these investigations give some ideas on how to proceed further and what might be expected.

The mathematics teacher operating for the first time in this new classroom mode will have understandable fears about order, control and confidence. However, there are plenty of examples in other areas of the established school curriculum where an instructional mode does not apply and where the teachers concerned operate with confidence and with no apparent lack of order. Once the pupils are involved and engrossed in an intriguing problem as already suggested, many of these fears will prove to be unfounded. If this initial hurdle is successfully negotiated further development of teaching style and pupils' thinking skills should follow.

TEACHING STRATEGIES

Much good advice on teaching problem-solving has been written over many years and a limited amount of research has been carried out into its implementation in the classroom. Polya's well-known set of heuristics[1] are principally addressed to the problem solver. A less well-known set of practical teaching techniques were devised by A.S. and E.H. Luchins in 1950.[2] These are briefly summarized, as follows.

- Having posed the starting point whether in written or verbal form, the teacher can orientate the pupil towards the problem by suggesting that difficulty may occur; this, if accepted, should provide motivation and challenge.
- Verbalization, dramatization and model construction should be encouraged.
- An atmosphere of question and discussion should be created, with praise for asking questions and avoidance of destructive critical comment.
- A heuristic method should be used, that is framing questions in such a way that the focus is changed or a new idea brought into the field.
- Inductive procedures should be encouraged, that is the type of procedures outlined in Chapter III, Mathematical Thinking.

On closer scrutiny of the context in which this advice was written it appears that these strategies refer to problem-solving within the traditional content-oriented curriculum, that is to problems, as opposed to exercises, associated with a particular item of mathematical knowledge. Nevertheless the teaching method implied has been found to be highly valid, in fact essential, for investigational work. However, the educational objectives are different in that the writers were concerned with enhancing the appreciation of some new extension to the learner's mathematical knowledge whereas our present concern is principally with developing mathematical thinking skills. What has been realized since this article was written is that thinking skills probably develop best in the context of mathematics which is firmly and confidently established, not that which has just been learnt.

A further stage, however, in the use of acquired thinking skills is to apply them to the acquisition of new mathematical knowledge; in this way the two aspects of the mathematical curriculum, knowledge and thinking, can complement each other.

RÔLE OF THE TEACHER

Investigational work implies a quite different rôle for the teacher than when an instructional or expository mode is used; the teacher is now a counsellor and overall co-ordinator. Various teaching techniques can be identified in addition to the heuristics already outlined in the previous section.

Once the starting point is established the teacher can

- organize ways of working, appropriate to specific investigations, individuals, pairs or groups, and encourage interaction between pupils;
- encourage the collection and recording of sufficient data;
- promote systematic and efficient methods of doing this;
- suggest thinking about modes of representation;
- provide appropriate resources and point to their existence;
- challenge pupils to make and justify assertions, to reflect on them and extend them;
- encourage recording and communication of observations and results, and the appreciation of the need for doing this;
- develop a system of class or group folders, individual files, and ongoing displays.

It cannot be stressed too strongly that at all times pupils should feel that they are responsible for initiating ideas and methods; they need time to develop their own understanding and insight. Decision-making should be in their hands and largely passed back to them when they ask questions. Observations and results which pupils put forward should not be explained by the teacher but readily received and accepted as neither right nor wrong; responses should take the form of 'Are you sure?', 'What happens if . . . ?', 'Are there any other possibilities?', or deflected to another pupil, 'Do you agree with that?'.

The good teacher can promote progress through an investigation by gentle probing and prompting at crucial moments. Pupils' data may appear to be incomplete, the organization of the data may not be systematic, they may just be in the respectable state of being stuck, or their written record may not be self-explanatory. It should be possible to extract pupils from such difficulties by judicious questions. The good teacher can suggest more data is obtained, diagrams are drawn, data arranged in rows or columns or tabulated, patterns looked for, and can ask for clarification of written notes and diagrams.

In a recent study Pirie[3] develops this teaching style in more detail, describing the teacher's rôle as an instigator, enabler, facilitator, listener, questioner, positive evaluator and observer.

CLASSROOM LAYOUT

Furniture is best deployed in U-shaped groups so that all pupils are facing, or sideways on, to some natural focus in the room, which may of course be the traditional blackboard. Such an arrangement encourages interaction between pupils, particularly within groups yet at the same time allowing for whole class discussion when appropriate. It is useful to have a spare chair with each group so the teacher can sit down when discussing progress; this enhances communication by reflecting equality of status in the development of ideas.

Resources (such as graph and grid papers, centicubes, rulers and scissors, etc.) should be easily and quickly accessible to all, with an understanding that the teacher need not always be consulted in respect of their use.

TIME CONSTRAINTS

With the pupils having a significant amount of responsibility for the manner in which ideas are developed it is unrealistic to prepare lessons with some notion of reaching a targeted stage in an investigation. Obviously, with experience of using the same starting point with different groups, there will come some realization of what might be expected. Planning should take the form of preparation of resources and an overall time scheme for different types of activity, consideration of the mathematical knowledge and type of thinking skills likely to be involved in a particular investigation; indeed the choice of investigation may hinge on these considerations.

Some pupils will often seem too occupied to be disturbed; nevertheless the established technique of always drawing a class or group together immediately before or after a break generally still applies. It will often be necessary to continue into another session.

Although, ideally, developing thinking skills should be an integral part of the whole mathematics curriculum and not a separate entity, there is a very good case to be made for having a sustained period of time at least once a term wholly devoted to investigative work, a week or even two weeks, to allow for a full enquiry with extensions into a particular starting point. This could be seen in National Curriculum terms as specifically aimed towards ATs 1 and 9, and to some extent 12 and 13.

REPRESENTATION

ATs 12 and 13 focus largely on data collection, recording, representation and interpretation techniques associated with the established statistical element of the mathematics curriculum. Investigation work uses many of these and frequently requires a more flexible approach; examination of the commentaries will show the regular use of tabulation to organize data. There is also a frequent need to invent some shorthand notation, such as ordered pairs, triples and so on to represent particular cases; see for example CUSTARD PIES (11).

RECORDING AND REPORTING

Recording means making rough notes, throughout the progress of an investigation and is a skill which should be fostered from the very beginning until it becomes routine. Rough sheets should be numbered and rudimentary organization encouraged. Then these notes should be such that they will be clearly understandable to their author when looking back and particularly when the time comes to communicate the findings in the form of a report.

The length of the resulting report will depend to some extent on the level of attainment but largely on the amount of time spent on the investigation. Pupils' and students' individual styles vary considerably. However, the report should be tidy and comprehensible to any person likely to read it, with some use of full sentences. It should contain

- the starting point,
- a description of how the investigation developed, making regular use of diagrams and tables,
- results and observations, and where appropriate some attempt at evaluating and explaining them,
- some outline suggestions of how the investigation might have been developed further, had more time been available,
- acknowledgements where appropriate to people or books where these have been consulted (this section of the report is more likely to be relevant to student investigations).

In the National Curriculum, recording is listed from level 2 upwards and reporting from level 4 upwards, in ATs 1 and 9.

ASSESSMENT

All assessment should be both diagnostic and prospective; diagnostic in so far as it determines the level of mathematical thinking both in National Curriculum terms and in terms of the thinking modes outlined in Chapter III, Mathematical Thinking; prospective in so far as it indicates what future assignments are appropriate for the pupil or student.

Generally, assessment will focus on the written report, but having worked alongside pupils/students in the development of an investigation the teacher will have a good idea of the level of thinking that underlies the production of the report. In addition, there is the quality of the report itself, the way in which it is organized and presented, to be considered.

It is probably always useful to attempt to grade a report under distinct headings and certainly essential to do this before attempting to judge it in its entirety. However, experience will lead eventually to an ability to judge and grade a report as an entity. This has been found to be an acceptable method, with plenty of evidence that teachers independently assessing the same report can show a remarkably good degree of agreement about the level achieved.

For the purpose of assessing all the reports written on the investigations in this book, a tabulated sheet was used for entering grades and written comments. Each sheet contained a check-list devised for grading under four equally-weighted headings.

1. Generalization: recognizing/extending patterns, generating examples, systematic classification, considering special cases, expressing relationships in some general form.
2. Explanation/proof: verbal explanations of observed patterns, making conjectures and empirically verifying them, use of counter examples and exceptions, formulating new questions, inductive argument, deductive proof.
3. Representation: appropriate use of diagrams, tables, charts, graphs, models, shorthand and symbolic notation.
4. Presentation: clarity, comprehensibility, overall structure, style and interest, effort, conclusions.

This check-list is clearly criterion-referenced; the statements of attainment for each level of the National Curriculum are also criteria and should enable fairly accurate assessment of the level of attainment instead of using grades.

Although the general principles of assessment within the National Curriculum are contained in the report of the Task Group on Assessment and Testing (DES 1988) full details have yet to be worked out. At Key Stage 4 there can be little doubt that the established methods of assessment now being developed within the GCSE context will be adopted and developed.

REFERENCES

1 Polya, G. (1945) *How To Solve It* Princeton, p. xvi.
2 Luchins, A.S. and E.H. (1950) New experimental attempts at preventing mechanization in problem-solving, *Journal of General Psychology*, 42.
3 Pirie, S. (1987) *Mathematics Investigations in Your Classroom* Macmillan.

V   THE NATIONAL CURRICULUM

In the preceding chapters there are frequent references to the Attainment Targets (ATs) of the National Curriculum. The more detailed analysis here attempts to show the overall links between the ATs and the investigations, which should enable teachers to identify an investigation suitable for use in connection with a particular mathematical topic at a particular level. However, it should be emphasized that particular investigations can be explored at all levels from a stated minimum and frequently are linked to more than one of ATs 2, 3, 5, 6, 7, 8, 10 and 11.

COMMENTS ON ATTAINMENT TARGETS

ATs 1 and 9: Using and applying mathematics

These two are very similar, the first concerned with number, algebra and measures, and the second with shape and space. All investigations are primarily aimed at these targets and give scope for work on one or other (or both) of them at an appropriate level.

ATs 12 and 13: Handling data

It has already been remarked in the section on Representation in Chapter IV that most investigations involve collecting, recording and representing data. They thus give scope for the parts of these targets which are not directly concerned with the statistical skills which form the essence of the higher levels in these ATs.

ATs 2 to 4: Number

There are many instances where levels within ATs 2 and 3 can be identified in the investigations, but there are few explicit links with AT4 which is concerned with estimation and approximation. Some number concepts are curiously absent from these ATs, for example number bases (perhaps implicit in AT2). Also, some references appear confusing, with equivalence of fractions apparently being level 6 of AT2 but level 4 of AT5.

AT5: Number/Algebra

The development of this AT is strongly exploratory in nature, leading from patterns to generalizations. Consequently, it is not surprising that many of the investigations are relevant to this particular Target.

ATs 6 and 7: Algebra

Use of notation, conventional or otherwise, is frequent in the investigations as is the idea of functional relationships. Coordinates (also listed in AT10) are sometimes employed; double-entry tabulation has much in common with coordinates and occurs in such as DIAGONAL (6), POCKET (8), CHEESE (9), CRAZY PAVING (14). Only older pupils and students are likely to employ the algebraic techniques which form the higher levels of these two ATs.

AT8: Measures

The context of this target is largely real-life problems; four of the investigations are directly relevant and all are concerned either wholly or partly with area and volume, THIRDING A SQUARE (16), FIGURE SEQUENCES (17), CUBE LINKS (30) and ROUND THE BEND (38).

Square counting is seen as level 4 whereas it is not until level 8 that area formulae are mentioned. It is reasonable to identify a stage, say at level 6, in between these which involves an understanding of congruence which is level 5 of AT10; such a level is used in these particular investigations. One or two investigations involve other types of measurement, for example BALANCING ACT (33) and CUSTARD PIES (11).

ATs 10 and 11: Shape and Space

A large proportion of the investigations have links with these ATs. Problems which start in a geometrical context are perhaps less abstract than those which are purely in the realm of number. Furthermore they can often involve use of number as well.

Three particular investigations, CUSTARD PIES (11), TREES (26) and DUOGONS (32) involve the use of networks which is level 5 of AT11. The idea that some networks are intrinsically the same, that is they can be distorted without altering their nature, provided the way in which the points are connected to each other by lines is unchanged (for example the London Underground map), does not receive explicit mention within the ATs. This is a primitive form of identity or congruence dependent only on points and lines (technically referred to as **topological equivalence**), and is estimated as level 3.

AT14: Handling data; probabilities

There is very little in the way of exploration of probability in the investigations; however for older pupils CHESSBOARD RECTANGLES (24) involves awareness of combinatorial ideas at level 6; such ideas could also be useful in ISOLATIONS (4) and RANDOM DOTS (10).

CROSS-REFERENCES BETWEEN INVESTIGATIONS AND ATS 2, 3, 5 TO 8, 10, 11

Investigation	Starting level	Number		Num./Alg.	Algebra		Measures	Shape and Space	
		AT2	AT3	AT5	AT6	AT7	AT8	AT10	AT11
1. CHAIN TRAPS	2		3	2	5				
2. STAIRCASE NUMBERS	2		3	2					
3. DELTOMINOES	1			6				1	
4. ISOLATIONS	1	1		6					
5. WALLS	2			6	5			2	3
6. DIAGONAL	3	6		5	3			3	6
7. POSTAGE	2		2	6					
8. POCKET	3	6		3	3				3
9. CHEESE	2			3				3	2
10. RANDOM DOTS	3							3	
11. CUSTARD PIES	3						5	4	3
12. DESIGNING FIGURES	3			4	4			3	
13. SUMS AND PRODUCTS	4		4	7	6		4		
14. CRAZY PAVING	4			5			4		
15. CLONES	6	6						6	6
16. THIRDING A SQUARE	4						4	5	4
17. FIGURE SEQUENCES	6	6					6	6	8
18. LIGHTHOUSES	6					7	6	10	6
19. GRACEFUL FIGURES	2		2	2				2	
20. COIN SHAPES	2			2				2	3
21. CHOP	3			3				3	

Cross-reference chart: Investigations and ATs (*cont.*)

Investigation	Starting level	Number AT2	Number AT3	Num./Alg. AT5	Algebra AT6	Algebra AT7	Measures AT8	Shape and Space AT10	Shape and Space AT11
22. PALINDROMIC SUMS	3	3	3						
23. NUMBER SPIRAL	3			3					
24. CHESSBOARD RECTANGLES	3							3	3
25. SPIROLATERALS	2			2		4			2
26. TREES	3				3				3
27. PAVING WITH DOMINOES	2							2	2
28. LOOPS	3			3	4			4	
29. MAGIC	3			3	7			3	
30. CUBE LINKS	4						4	4	
31. DIFFERENCES	4	4	4						
32. DUOGONS	3			4					3
33. BALANCING ACT	5	5					5		
34. QUADRUPLES	5				5	7		7	
35. DESIGN	5					5		5	
36. VECTOR PATHS	4					4			4
37. PIVOT POINTS	4							4	4
38. ROUND THE BEND	6						6	8	
39. SQUARE SPIRAL	4		6					4	6
40. FRACTION CHALLENGE	6	6	6						6

The chart above shows firstly the overall minimum level for each investigation, then the appropriate minimum level for each relevant AT. Inevitably tackling some aspects of the investigations may imply a higher level within an AT than that stated. Because of the universality of ATs 1, 9, 12 and 13 these have been omitted; ATs 4 and 14 have also been omitted because of their limited relevance.

The information provided by this chart is a very rudimentary checklist and should be used only in conjunction with the commentaries on the investigations.

INVESTIGATIONS: FULL COMMENTARIES

1   CHAIN TRAPS

STATEMENT OF PROBLEM

Here is a number trap! You enter it by thinking of a number; try a small one to begin with, less than 10 say. Follow the instructions, which tell you what to do with your number. If you can make your number into 1 you can escape from the trap.

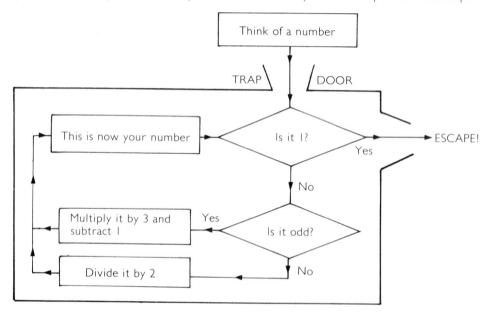

Obviously, if you choose to enter the trap with 1 then you can escape straightaway! But, if you enter with 6 then you change it along the chain

$$6 \rightarrow 3 \rightarrow 8 \rightarrow 4 \rightarrow 2 \rightarrow 1: \text{ESCAPE!}$$

Try other numbers and make a note of the number of changes you have to make before escaping. If the numbers become too big use a calculator.

Can you always escape?

Design some other traps by varying the changing rules, or even the conditions for escape.

COMMENTARY

National Curriculum

This investigation is for level 2 or 3 upwards; it affords a lot of experience within AT5, using patterns and **sequences** and making generalizations. Level 2 of AT5 refers to odd/even numbers; level 3 of AT3 mentions the relevant arithmetical operations and use of a calculator; for computing to be used level 5 is more appropriate (AT5).

There is also plenty of experience in creating analogous problems.

Materials

Squared paper helps to organize the recording of data. The use of a calculator is straightforward and very helpful in speeding up the production of data; it also ensures accuracy when numbers are large and chains are long.

GENERAL COMMENTS

The close relationship with the classic chain problem

if odd multiply by 3 and *add* 1, if even divide by 2

will be noted; this classic problem dates back at least into the last century. Whether or not the chain of numbers always reaches 1 whatever the starting number is as yet unresolved in spite of the activities of many minds over many years but there is little doubt that it always does. This may seem rather daunting but the purpose of this investigation is not to attempt to solve the so far unsolved but to find patterns in the way the chains behave and possibly explain them.

The slight variation from the classic problem in the statement of this investigation, that of changing 'add 1' to '*subtract* 1' makes a significant difference. Some chains instead of always reaching 1 fall into repetitive cycles such as

$5 \rightarrow 14 \rightarrow 7 \rightarrow 20 \rightarrow 10 \rightarrow 5$

which has five links; besides 5, 7 and 10 starting numbers 9, 13 and 14 and many others also are trapped by this cycle. Another longer cycle with 18 links is entered if 17 is the starting number and this cycle inevitably traps many other numbers.

Strictly speaking, starting numbers which reach 1 are also effectively trapped by a cycle, namely $1 \rightarrow 2 \rightarrow 1$, but this cycle contains the important escape number. What makes the classic problem particularly intriguing is that precisely *one* cycle traps all numbers and that cycle contains 1.

The problem is strikingly simple to program and use of a computer for the number crunching greatly enlarges the possible horizons of the enquiry.

INITIAL APPROACHES

Pairs of pupils working together is stimulating at the start. Each can design a trap from which the other tries to escape; in the interest of fair play cross-checking the correctness of chains will also be ensured. Younger pupils possibly should be encouraged to use simpler conditions than those in the stated problem such as

if odd add 3, if even divide by 2;

for this they should discover that you only get trapped if you enter with a multiple of 3, and possibly attempt an explanation of this observation.

Some simple structural points should also emerge from discussion at this stage of the investigation if the 'if even divide by 2' condition is retained:

- any power of 2 will not be trapped;
- the condition for odd must be such as to convert the odd number to an even one;
- if an entering number is not trapped then neither is its double.

Recording of chains, in rows or columns on squared paper, though not necessary at the start of the exploration, soon becomes essential.

CHANGING CONDITIONS

Changing the 'subtract 1' condition in the stated problem provides an interesting development. Some observations which have emerged in considering conditions of the form 'if odd multiply by 3 and add n, if even divide by 2' for various values of n are as follows.

$n = 3$: All numbers except powers of 2 get trapped by the cycle

$$3 \rightarrow 12 \rightarrow 6 \rightarrow 3.$$

However, if 1 is not the escape condition it also is trapped by this cycle and so we have a variant of the classic problem in that all numbers appear to converge to 3, which makes 'Is your number 3?' a universal escape condition.

$n = 5$: At least three different cycles occur, starting with 5 (three links), 19 (eight links), and 23 (eight links).

$n = 7$: At least two cycles occur, starting with 5 (six links) and 7 (three links).

$n = 9$: This is similar to $n = 3$; no numbers appear to be trapped if the escape condition is changed to 'Is your number 9?'.

Similar observations occur for other values of n but to explore these in greater depth, computer power is appropriate (see the later section on computing).

The chains with $n=41$ look promisingly at first sight to be a repeat of the classic problem. All numbers appear to converge on 1, although some of the chains are particularly long, with 5, for example, taking 45 links to reach 1. Unfortunately when 41 is considered it cycles

$$41 \rightarrow 164 \rightarrow 82 \rightarrow 41$$

and likewise multiples of 41 follow suit. This is reminiscent of that remarkable 'formula' for prime numbers $n^2 - n + 41$, which is prime for all values of n from 1 to 40. The chains for $n=61$ follow similar lines to that for $n=41$.

EXTRA CONDITIONS

Other variations to the conditions can be tried but there are problems.

If odd multiply by 5 and add 1, if even divide by 2

results in some starting numbers, for example 7 and 9, increasing apparently without limit. One suggestion put forward to overcome this is to introduce an extra condition 'if a multiple of 3 divide by 3' and with this included it appears that, like the classic problem, all numbers again converge to 1. This phenomenon of spiralling out of control shows how subtly balanced are the two conditions of the classic problem. All variations in the conditions have therefore to be similarly balanced if this phenomenon is to be avoided.

Another variation suggested which has 1 as a universal escape is

'if odd multiply by 2 and add 2, if even divide by 2'.

With this rather simple case it is not difficult to explain why all starting numbers converge to 1.

ANALYSING THE CLASSIC PROBLEM

Some pupils may concentrate on examining the classic problem in detail and again computer assistance is rewarding. One way of doing this is to look at the length of chains to 1. Some significant observations are listed below.

1. The length of chain varies dramatically and apparently erratically; for the starting numbers 1 to 26 the maximum number of links in the chain to 1 is 23, but 27 has no less than 111 links! 54 therefore has 112 links but curiously so does 55; perhaps the variation of length is not so erratic after all.

2. It is then a long way to go before a starting number is found with more than 200 links, namely 2463 with 208 links. This slow growth in the length of chain possibly adds confidence to the hypothesis that all chains converge to 1.

3. Not only do 54 and 55 have the same number of links but numerous other pairs of adjacent numbers starting with 12 and 13, 14 and 15, 18 and 19, 20 and 21 and so on. This phenomenon calls for an explanation. In the case of pairs of numbers it can soon be noticed that for some pairs their chains converge after only three links, for example

$$12 \rightarrow 6 \rightarrow 3 \rightarrow 10 \ . \ . \ .$$
$$13 \rightarrow 40 \rightarrow 20 \rightarrow 10 \ . \ . \ .$$

This also happens with $(20, 21)$, $(28, 29)$, $(36, 37)$, . . . which are all of the form $(8n+4, 8n+5)$ where n is an integer. Feeding two such numbers into the trap gives

$$8n+4 \rightarrow 4n+2 \rightarrow 2n+1 \rightarrow 6n+4$$

and

$$8n+5 \rightarrow 24n+16 \rightarrow 12n+8 \rightarrow 6n+4.$$

In functional notation, let

$$f(x) = 3x+1 \quad \text{and} \quad g(x) = \frac{x}{2};$$

then

$$fg^2(8n+4) = 6n+4 \quad \text{and} \quad g^2f(8n+5) = 6n+4.$$

4. Other pairs such as $(14, 15)$ become another pair $(22, 23)$ after two links and eventually to a pair of the form $(8n+4, 8n+5)$ and so converge after three more links.

How are triples of adjacent numbers with the same number of links explained, for example 28, 29 and 30?

COMPUTING

Programming the problem has already been mentioned. The key line in a simple program to generate chains for the classic problem is of the form (in BASIC):

IF N/2 = INT(N/2) THEN N = N/2 ELSE N = 3*N+1

This is enclosed in a REPEAT UNTIL loop with appropriate PRINT and INPUT commands added. The addition of a counter variable is useful.

If traps which result in cycles are to be explored then each chain needs to be kept in a list and each new number checked against the list.

EXTENSIONS

Most of the variations discussed in the last section are fairly close to the original form of the problem. Chains of numbers can, of course, be created in numerous ways and in this sense there is no limit to possible extensions. However these need to have a focus, the focus of the present investigation being on chains which end in the same number regardless of the starting number. Other chains could be those that always return to the starting number, that is all numbers form cycles.

With older pupils and students one concept which can be sensibly explored is that of a **limit**; a chain formed by continual application of the loop

add 4, divide by 5

has 1 as a limit for all starting numbers; the escape condition in such a chain could be a test for nearness to 1. A slightly more intriguing loop is

divide 8 by your number, add 2.

This type of extension however deserves to be treated as a separate investigation.

2 STAIRCASE NUMBERS

STATEMENT OF PROBLEM

All numbers are interesting! For this investigation we start with 15. 15 is interesting because it can be expressed as the sum of consecutive integers (a staircase) in three different ways.

Can you find other such interesting numbers? Start perhaps by looking for integers which have two staircase forms, for example 9 which is both 2+3+4 and 4+5.

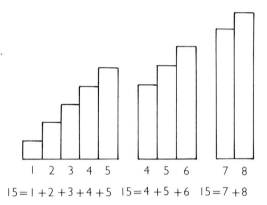

1 2 3 4 5   4 5 6   7 8

15=1+2+3+4+5   15=4+5+6   15=7+8

COMMENTARY

National Curriculum

This investigation is for level 2 upwards; like CHAIN TRAPS (1) it is principally concerned with AT5. Clearly it would be quite a suitable idea to explore whilst children are learning their **number bonds**. At higher levels the investigation offers good opportunities for generalization, plausible conjectures and in some instances, proof.

Materials

Cuisenaire rods or other equivalent structural apparatus are appropriate; 15 can be made from three different staircases of rods. Otherwise squared paper should be used for drawing staircases and tabulating data.

INITIAL APPROACHES

This is an investigation where, initially, an indirect attack appears more productive than a direct attack. The direct attack is to take a number, say 22, and try to write it in the form of a consecutive integer sum; 4+5+6+7 is not immediately obvious and even more-able pupils find it difficult to formulate methods for getting *to* such a sum *from* the number. After some mulling, using this direct approach to start with, an indirect approach begins to emerge – by listing the numbers which are the sums of two, three, four, five consecutive integers and so on.

Two:	Three:	Four:	Five:
$1+2=3$	$1+2+3=6$	$1+2+3+4=10$	$1+2+3+4+5=15$
$2+3=5$	$2+3+4=9$	$2+3+4+5=14$	$2+3+4+5+6=20$
$3+4=7$	$3+4+5=12$	$3+4+5+6=18$	$3+4+5+6+7=25$
$4+5=9$	$4+5+6=15$	$4+5+6+7=22$	$4+5+6+7+8=30$
$5+6=11$	$5+6+7=18$	$5+6+7+8=26$	$5+6+7+8+9=35$
etc.	etc.	etc.	etc.

From such data we can pick out that 18, like 9, is also a two staircase number.

Pupils often miss some of the longer staircase forms, for example finding $22+23$, $14+15+16$ and $7+8+9+10+11$ for 45 but omitting $5+6+7+8+9+10$ and $1+2+3+4+5+6+7+8+9$.

GENERALIZING STAIRCASE SEQUENCES

These sequences of sums can be generalized (in a variety of ways) for example:

1st	2nd	3rd	4th	5th	...	nth number
3	5	7	9	11	...	$2n+1$
6	9	12	15	18	...	$3n+3$
10	14	18	22	26	...	$4n+6$
15	20	25	30	35	...	$5n+10$

The patterns observable in these sequences should be explained and in particular such attempts at explanation may well lead to the early stages of a method for determining staircase forms for a given number. For example, when there is an odd number of stairs, the staircase number is a multiple of 3, 5, 7 and so on. Thus if an integer is a multiple of 3 it must have a three-stair form.

For an even number of stairs, a method is not so easy to formulate since, for example, four consecutive integers do not add up to a multiple of four; it is likely that a variety of effectively equivalent methods will be suggested.

TABULATING THE DATA

It should now be possible to tabulate the integers according to the number of ways they can be represented as staircases. As usual with such tables a lot of effort is involved which is possibly best shared. The following table shows all the integers up to 16 and from then on only some of the succeeding integers in each of the columns. The larger integers have been included to indicate more strongly some of the patterns present. Since most numbers between 1 and 100 are in the first four columns, when pupils construct such a table entries in the other columns will be few and need careful checking.

Number of ways each integer in the table can be expressed as a staircase

0	1	2	3	4	5	6	7
1	3	9	15	81	45		105
2	5	18	21	162	63		135
4	6	25	27	225	75		165
8	7	36	30		90		189
16	10	49	33		99		
32	11	50	35		117		
64	12	72	39				
	13	98	42				
	14						

CONJECTURES

The table above bristles with plausible conjectures, some of which are put forward often before tabulation is attempted. For example:

- all powers of 2 have no staircase form;
- all primes have only one staircase form;
- if an integer has n staircase forms then so does twice the integer. (This is structurally similar to the first conjecture.)

The table also prompts a lot of questions.

- What about integers with six staircase forms? Are there none? If not, what is the smallest? (This set of questions could apply at an earlier stage to the column headed 4.)
- Where do squares occur in the table and why?
- What are the characteristics of integers with a large number of staircase forms?

Answering and attempting to explain these conjectures and questions provokes a variety of ideas and, with able pupils and students, can lead eventually to a satisfactory complete formulation for determining the total number of staircase forms for a given number. Some stages on the way to this formulation, which extend the three conjectures, are shown in the following table.

Form of integer*	2^p	m	$m \times 2^p$	m^2	m^3	$m \times n$	$m^2 \times n$
Number of staircase forms	0	1	1	2	3	3	5

*p is any positive integer and m and n are unequal prime numbers (excluding 2).

EXTENSIONS

Other types of staircase are the most obvious extension, for example in steps of 2 ($1 + 3 = 4$ and so on). Formulating a method for determining such staircase forms, curiously, is easier than the original unit-step problem since two stairs give a multiple of 2, three stairs a multiple of 3 and so on.

With steps of 3 however there are similar problems to the unit-step case; two-way correspondences between the two cases can be set up, for example for 30

$9 + 10 + 11 \longleftrightarrow 7 + 10 + 13$
$6 + 7 + 8 + 9 \longleftrightarrow 3 + 6 + 9 + 12$

This is a stretching transformation keeping the centre fixed, with 7.5 being the 'centre' in the second case. However $4 + 5 + 6 + 7 + 8 = 30$ has no corresponding three-step case because 0 would be involved.

It seems reasonable to assert that to every three-step staircase form of an integer there corresponds a one-step staircase form, though not always vice versa. Would the same be true for two-step staircase forms?

3 DELTOMINOES

Statement of Problem

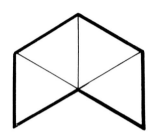

A deltomino is a figure made by joining equilateral triangles edge to edge. The diagram shows a 4-deltomino.

Can you find some different 4-deltominoes?

You can explore the world of deltominoes in various ways. For example:

- try to find all the different 5-deltominoes;
- design a deltomino and try forming a tiling with it;
- design a deltomino and by folding it along its edges see whether it will fold up to form a shape with triangular faces.

Commentary

National Curriculum

This investigation is for level 1 upwards and should contribute significantly to achieving aspects of AT10. Constructing deltahedra is level 4 and the enumeration and **tessellation** of the higher-order deltominoes is level 5 or 6.

Materials

At the earlier levels, pupils will prefer to work with gummed triangles and/or templates. Later, obviously, triangled (isometric) paper can be of considerable assistance for drawing, enumerating and tessellating deltominoes, and for constructing **nets** for deltahedra.

General Comments

The investigation is clearly an extension of the world of the square-based polyominoes, pentominoes and hexominoes in particular. That there are 12 distinct pentominoes is widely known as there are many puzzles and games based on them; they are strangely fascinating. Perhaps less well known is that they can all be used individually to tile a surface without overlapping, and that eight of the twelve can be folded to form a cubical box without a lid. The ideas suggested in the statement of the problem invite exploration of the deltomino world in a similar manner.

ENUMERATION: FIRST STAGES

Enumeration of the small sets of deltominoes is quite straightforward up to 5-deltominoes. On the other hand, finding the twelve distinct 6-deltominoes is quite a challenge and definitely needs a systematic approach. Early decisions have to be made about what counts as different; mirror and rotatable images are usually discounted in order to reduce the total of different cases. Square-ominoes are usually drawn with horizontal/vertical orientation and duplicates are fairly easy to pick out by eye; however, even with the 5-deltominoes, problems of avoiding duplication are not quite as simple because of the inclined orientation of some potential duplicates. Young pupils may need to cut each deltomino out for direct comparison with each of the others; older pupils try various methods of isolating different types within each deltomino set and some of these methods are discussed in the next two sections.

SYSTEMATIC ENUMERATION

This is quite feasible up to 7-deltominoes and for some careful enthusiasts up to 8. Beyond that point it would appear vital to establish some sort of generalized recurrence relationship between each set of deltominoes. An ingenious though rather complex and limited method is described by Ellard[1] which confirms the figure of 66 for 8-deltominoes (obtained by a student some ten years previously) and confidently projects numbers for the next 4 sets, for example 3342 12-deltominoes!

A popular method of classification is by the length of the spine, the longest linear arrangement of triangles contained in the deltomino. Of the four 5-deltominoes shown below one has a maximum spine of 5 triangles, two of 4 and one of 3. The spine can occur in any one of three directions.

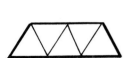 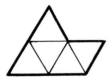 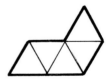

In the 6-deltomino set there is clearly just one with a spine of length 6; those with a spine of length 5 can be found by adding one triangle to such a spine without lengthening it, three distinct cases being possible. Two triangles are then added to a spine of length 4 and so on.

The process of adding one triangle to a 5-deltomino to form a 6-deltomino can be used as an alternative method of enumeration. There are seven possible positions where an additional triangle can be attached to each of the four 5-deltominoes; this gives a set of 28 potential 6-deltominoes from which duplicates are eliminated, some easily spotted as the set is being drawn. Other duplicates are difficult to distinguish and shading alternate triangles is some help.

ANALYSING THE SEQUENCE OF DELTOMINOES

It seems that the number sequence for all possible n-deltominoes which goes

1, 1, 1, 3, 4, 12, 24, 66, 160, 448, 1186, 3342, . . .

(the last four being the 'Ellard' numbers mentioned in the previous section) has no easy formulation, which is not unusual for sequences produced from geometrical specifications.

On the other hand, some subsets of the sequence appear to be more definable.

1. Deltominoes in which all but one of the triangles are on the spine have been found to give rise to a simple sequence as shown below.

Triangles (n)	4	5	6	7	8	9	10	11	12
Maximum spine length ($n-1$)	3	4	5	6	7	8	9	10	11
Number of deltominoes	2	2	3	3	4	4	5	5	6

2. Those which have all but two of the triangles on the spine also give rise to sequences which are capable of generalization and eventual formulation. Problems arise in the early cases of 5- and 6-deltominoes, for example attaching two extra triangles to a spine of length 3 frequently creates an alternative spine of length 3; such problems disappear with 7-deltominoes and upwards. A plausible formulation for deltominoes of this type is

for $n > 5$ and odd, $\frac{1}{4}(n^2 - 5)$;
for $n > 6$ and even, $\frac{1}{4}(n^2 - 4)$.

3. A smaller subset consists of those 'reflectable' deltominoes which have at least one line of symmetry. These appear to form a sequence of pairs.

Triangles (n)	1	2	3	4	5	6	7	8	9
Number of deltominoes	1	1	1	2	2	5	5	12	12

Whether or not the pattern continues remains a matter for speculation. It would appear reasonable to suggest that an even number of triangles is easier to form into a reflectable deltomino than an odd number, but no obvious matching between, for example, the reflectable 6- and 7-deltominoes has been found.

4. In an attempt to distinguish between the deltominoes of a particular set some experimenters have focused on the way the triangles appear in the deltomino; some triangles are totally enclosed, others have one 'free' edge and the rest have two free edges. In the diagram are three of the 7-deltominoes with the triangles labelled 0, 1 or 2 accordingly.

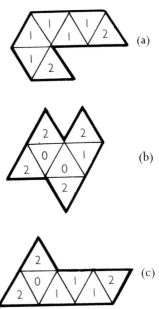

It is soon remarked, with good reason, that the number of triangles with 2 free edges is two more than those with none. Another interesting subset has been isolated, those deltominoes which have no enclosed triangles and are in the form of a chain of triangles. Deltomino (a) in the diagram is a chain.

The following sequence has been found.

Triangles (n)	1	2	3	4	5	6	7	8
Number of chain deltominoes	1	1	1	2	3	5	8	15

The hexagonal 6-deltomino has been excluded as being a loop rather than a chain, but this may be special pleading on the part of those who fancied that a Fibonacci-type sequence (in which each number is the sum of the two preceding ones) was in the offing; unfortunately the finding of fifteen 8-deltomino chains upsets this notion. Once again it is a matter for speculation as to how this sequence develops.

TESSELLATING DELTOMINOES

Looking at deltominoes as tessellating tiles is quite independent of the various processes of enumeration so far explored. Deltominoes offer a much wider possibility for pattern-making than the square-based polyominoes; organized repeating patterns with three- and sixfold symmetry are possible as well as those with two- and fourfold symmetry.

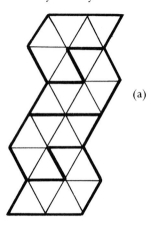

(a)

Young pupils should be encouraged to focus on the 1-, 2- and 3-deltominoes using them either singly or in combination. A surprisingly large number of different patterns can be created on triangled paper with them; use of colouring adds strongly to the effect.

With some of the higher-order deltominoes it is not always easy to establish an organized tessellation in the conventional sense of having no overlaps and gaps; furthermore it seems highly probable that some cannot be formed into such a pattern. A typical strategy which can lead to success is to fit two (or three) together to form a less complex figure, or one which will more obviously form a tiling. Example (a) in the diagram shows such a strategy applied to a 6-deltomino; deltomino (b) however has frustrated several tilers.

(b)

CONSTRUCTING DELTAHEDRA

Viewing deltominoes as potential nets for polyhedra with triangular faces (deltahedra) is another approach largely independent of the preceding ones. The single 3-deltomino can easily be seen to fold to form a triangular 'cup', and pupils in any doubt about this should perform it physically. The extra triangle in the 4-deltominoes will form a lid for this cup (that is form a tetrahedron) in two of the three cases.

If two tetrahedra are placed face to face the deltahedron formed has six exterior triangular faces; the 6-deltomino shown will fold to form this deltahedron. How many other 6-deltominoes can be found which will do likewise? How many 5-deltominoes will form this deltahedron with one face missing?

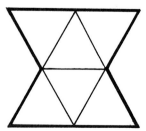

The regular octahedron can be formed from some of the 8-deltominoes; in the process of exploring such possibilities another distinct 8-faced deltahedron has been found; this prompts the question of how many 10-faced deltahedra there might be. The regular icosahedron is a twenty-faced deltahedra; it has been hypothesized that no 18-faced deltahedron exists!

EXTENSIONS

Having found the problem of enumeration of deltominoes absorbing it is not unusual to refer back to the square-based polyominoes which was the original source of this investigation. A fairly well known result is that there are 35 hexominoes, of which 11 form cubical nets, but how many heptominoes are there?

The other regular tessellating figure is the regular hexagon and this is worthy of some attention. Hexomino would seem an eminently suitable name for figures made by joining regular hexagons edge to edge but it has already been claimed. For fairly obvious reasons, these types of figure increase at a greater rate than either deltominoes or square-based polyominoes.

REFERENCES

1 Ellard, D. (1982) Polyiamond enumeration, *Mathematical Gazette* December, 310–4.

4 ISOLATIONS

STATEMENT OF PROBLEM

Here is a chain of the first four
whole numbers.

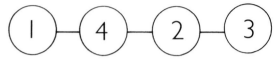

In this chain I and 4 are **isolated** because neither has its normal numbers next to it;
2 and 3 are not isolated.

Can you make a chain of the same four numbers so that they are all isolated?
Experiment with different sets of numbers and different sorts of chains.

COMMENTARY

National Curriculum

Level 1 of AT2 refers to ordering numbers up to 10 and this is sufficient to start this
investigation. For isolated chains of five numbers a systematic approach is necessary which
implies level 3 (AT1).

Materials

With young pupils a visual aid in the form of four cubes (easy to handle) or counters with the
numbers on them is helpful. Some children also prefer working with letters rather than
numbers.

Programming a computer to determine comprehensively all completely isolated chains of
seven numbers would be within the capabilities only of older pupils and students with
established computing skills.

GENERAL COMMENTS

When more than five or six numbers are used the number of cases to be investigated increases
considerably. Consequently it may be better to concentrate on varying the arrangements of
the numbers (see the Extensions section). For six numbers in a single chain there are 720
($=6!$) arrangements of which 90 have all six numbers completely isolated and for seven
numbers the corresponding figures are 5040 and 646.

INITIAL APPROACHES

With young pupils, start by reducing the problem to three numbers. They will soon notice that it is impossible to get all the numbers isolated and possibly be able to give reasons verbally. This gives impetus to go back to the four-number problem and to find with a little surprise that completely isolated chains are possible, discovering 2413 and 3142 without a great deal of difficulty; some may remark that these are really the same, each being the reverse of the other. Another suggestion may be 'keep the odds together and the evens together'.

ANALYSING THE FOUR-NUMBER CASE

It is possible to do this in detail looking at both partially and completely isolated chains; the arrangement 1243 has no numbers isolated whereas 1432 has just the number 1 isolated.

Of the 24 arrangements of the four numbers a detailed analysis could be presented in tabulated form as shown below.

Isolated numbers	Number of cases	Examples
0	8	1234
1	4	1432*
2	10	1342*
3	0	
4	2	2413* and 3142*
	Total 24	

*Isolated digits are shown in bold.

A few possible observations to explain are:

- if only one number is isolated then it is either 1 or 4 in an end position;
- if only two numbers are isolated then it cannot be the middle two;
- if a number next to an end is isolated then so is the end one (which helps to explain the previous two observations).

Pupils should be challenged to explain why it is not possible to isolate just three of the numbers.

CHAINS OF FIVE NUMBERS

With five numbers perhaps it is wise to concentrate at first on completely isolated chains. A systematic approach is vital and care must be taken not to overlook repeated solutions. One system frequently used is to start the chain with 1, then 2 and so on noting that every case found can immediately be reversed as already mentioned above.

There are 14 solutions and analysis of their interrelationship helps in deciding whether or not the set is complete. For example an older pupil produced the following diagram.

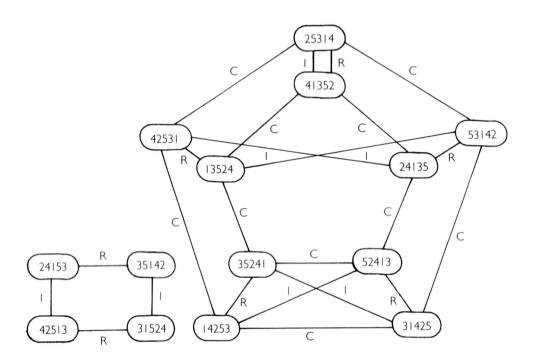

The diagram is based on the following three rules which produce related chains.

1. All completely isolated chains are reversible; these are connected in the diagram by a line labelled R.
2. If a completely isolated chain has two non-adjacent numbers at either end then it can be rotated cyclically, e.g. 13524 → 35241 and vice versa; these are connected by a line labelled C.
3. Any completely isolated chain can be inverted by subtracting all the numbers from 6, e.g. 24135 → 42531 and vice versa; these are connected by a line labelled I.

In the diagram there are only two fundamentally different chains of the numbers 1–5 from which all the others are derived.

FURTHER DEVELOPMENT

Tabulation of the number of completely isolated chains starting with specific numbers shows the beginnings of some structural connections between subsequent sets.

A computer program was used to reduce the number-crunching labour for the 1–6 and 1–7 cases. The program produced the desired results by first creating different arrangements of numbers, secondly checking each arrangement for complete isolation and thirdly checking that any new completely isolated chain differed from those previously found.

Numbers of completely isolated chains classified by chain length and starting number

Chains	Starting number							Total
	1	2	3	4	5	6	7	
1–4	0	1	1	0				2
1–5	2	3	4	3	2			14
1–6	12	15	18	18	15	12		90
1–7	78	93	100	104	100	93	78	646

One structural link connects the number of chains beginning with 1 from previous results. For example in the 1–7 case the number of completely isolated chains beginning with 1 is, by symmetry, the same as the number beginning with 7, and the latter must consist of 7 followed by all the completely isolated chains of 1–6 *except* those beginning with 6. So we have $(12+15+18+18+15=)$ 78.

Other connections are arguable, particularly if the various elements in the table are split up into those which are rotatable (according to rule 2 of the previous section) and those which are not.

EXTENSIONS

For younger pupils it is likely to be more fruitful to change the linking between the numbers than to follow the development discussed in the previous section. In discussing other possible linkings, the first suggestion is invariably to join up both ends of the chain and consider arrangements formed by the vertices of a square, pentagon and so on. The number of complete isolations is reduced to zero in the 1–4 case and 10 in the 1–5 case.

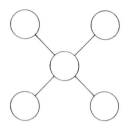

A star shape is another frequent suggestion as shown, but then soon ruled out as the central number is next to all the others.

Another arrangement worthy of consideration is the egg box for the numbers 1–6 as shown. The extra links reduce the number of possible completely isolated cases dramatically from the chain case.

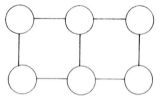

Redefining 'isolation'

Changing the definition of isolation is also a possibility, for example a difference of more than 2 between adjacent numbers; the arrangement 362514 obeys such a definition.

This also reduces the numbers of cases very dramatically; the number of completely isolated chains of 1–6 is reduced to two (one already mentioned above) and of 1–7 to 32. The first case is not difficult for any pupil to tackle and the second is achievable once some or all of the rules of the section on chains of five numbers are employed.

5 WALLS

STATEMENT OF PROBLEM

A brick is 2 units long, I unit wide and I unit high.

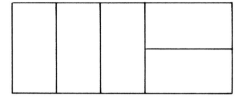

The figure shows a wall 2 units high, 5 units long and I unit thick built with five bricks.

Can you find some different designs for walls of the same size?

Analyse similar problems for different sizes of wall and, possibly, with different sizes of brick.

COMMENTARY

National Curriculum

This investigation is for level 2 upwards; in AT10 level 2 refers to creating patterns with 2-D shapes. It also uses and gives considerable experience of **reflective symmetry** which is level 3 (AT11). Some of the algebraic notation used for analysing the links between walls of different lengths implies at least level 5 or 6 (AT5).

Materials

Young pupils can build walls with number rods and then draw them; squared paper is advisable for drawing walls.

GENERAL COMMENTS

All walls in the investigation are assumed to be one unit thick, the third dimension being included in the statement of the problem to make it realistic.

The investigation is a development of one of the lesser-known ways of introducing the **Fibonacci** sequence,

1, 1, 2, 3, 5, 8, 13, 21, . . .

A wall which is 5 units long can be either a wall 3 units long extended by two horizontal bricks, or a wall 4 units long extended by one vertical brick.

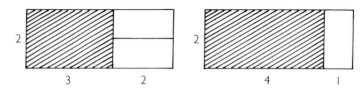

The number of different 5-brick walls is therefore the sum of the numbers of different 3- and 4-brick walls; the numbers of walls of lengths 1 and 2 are clearly one and two and so the sequence is formed.

However, this argument is far too subtle for many pupils and students to understand when presented to them in a demonstrative mode; it has been found much better to explore the problem from scratch and build up to and beyond the Fibonacci sequence.

INITIAL APPROACHES

Drawing as many different walls of length 5 as possible on squared paper is a good starting point. This exercise provides a good example of finding all cases to fit a given specification and, more importantly, knowing that the set is complete. If pupils find difficulty in resolving this task, then the length can be reduced to 4 or 3. Once the initial task is resolved then walls 6 units long can be considered and the Fibonacci sequence should begin to emerge. This may, of course, be recognized if pupils are already familiar with it but the reason why it occurs will not be clear until after quite a lot more exploration.

For some time it is necessary to enumerate all possible examples of a set by systematically organized drawings; eventually it is possible to argue more abstractly about the relationships involved in extending walls of known complete sets.

SYMMETRICAL AND NON-SYMMETRICAL WALLS

One aspect to consider when trying to determine whether or not a particular set is complete is to distinguish between symmetrical and non-symmetrical cases. Of the three walls of length 5 shown below, the first has a vertical line of symmetry and the other two are asymmetrical but form a mirror image pair; all sets should therefore be checked to see that every asymmetrical wall also has its mirror image in the set.

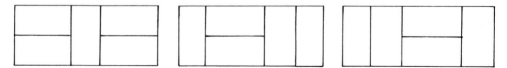

Splitting the sets up accordingly provides more information for speculation.

Length of wall (units)	1	2	3	4	5	6	7	8	9	10
Asymmetrical types			2	2	6	8	18	26	50	76
Symmetrical types	1	2	1	3	2	5	3	8	5	13
Total	1	2	3	5	8	13	21	34	55	89

The numbers for walls of the longer lengths are included here to show possible patterns more clearly, although it is not expected that pupils will in general venture beyond wall lengths of 8 units.

The numbers of symmetrical types of wall are also numbers in the Fibonacci sequence but in an oscillating fashion. Further, splitting up into odd and even lengths gives

symmetrical walls of odd length 1, 1, 2, 3, 5, . . .
symmetrical walls of even length 2, 3, 5, 8, 13, . . .

There would appear to be a connection between the number of symmetrical walls of a given length and the total number of walls of a shorter length. If s_n and t_n are the number of symmetrical walls and the total number of walls, respectively of length n then

$$s_1 = s_3 = t_1, \ s_5 = t_2, \ s_7 = t_3, \ s_9 = t_4, \ \cdot \ \cdot \ \cdot$$

and

$$s_2 = t_2, \ s_4 = t_3, \ s_6 = t_4, \ s_8 = t_5, \ s_{10} = t_6, \ \cdot \ \cdot \ \cdot$$

which generally suggests

$$s_{2n+1} = t_n \quad \text{and} \quad s_{2n} = t_{n+1}.$$

An explanation of the first of these two observations frequently put forward is as follows. All symmetrical walls of odd length have a vertical brick in the centre as the diagram shows. The left-hand shaded part can be any of the walls of length n and the right-hand shaded part is the mirror image of the left-hand part.

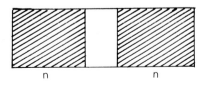

The second observation can be similarly but not so easily explained.

AN ALTERNATIVE APPROACH

Another approach to ensuring completeness of the sets is to envisage the walls constructed from two basic 'blocks' A and B as shown. The eight walls of length 5 units can then be enumerated as follows:

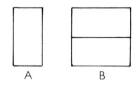

A   B

5 A blocks	AAAAA			
3 A blocks + 1 B block	AAAB	AABA	ABAA	BAAA
1 A block + 2 B blocks	ABB	BAB	BBA	

This method can then be applied to other lengths of walls and a tabulation made as follows.

Length of wall (units)	1	2	3	4	5	6	7	8	9	10
Walls with only A blocks	1	1	1	1	1	1	1	1	1	1
Walls containing 1 B block		1	2	3	4	5	6	7	8	9
Walls containing 2 B blocks				1	3	6	10	15	21	28
Walls containing 3 B blocks						1	4	10	20	35
Walls containing 4 B blocks								1	5	15
Walls containing 5 B blocks										1
Total	1	2	3	5	8	13	21	34	55	89

This table contains some recognizable sequences; furthermore the central section of it when read diagonally from top-left to bottom-right gives the successive rows 1 1, 1 2 1, 1 3 3 1, . . . of Pascal's triangle.

Why should each number in the table be the sum of the number to its left and the number diagonally above to the left of that, for example $21 = 15 + 6$, $35 = 20 + 15$, . . . ?

EXTENSIONS

Some possible extensions are suggested in the statement of the problem; two of them are considered in more detail here.

Higher walls with larger bricks

The simpler extensions are to walls 3 units high using bricks 3 by 1 by 1, and to 4 units high using bricks 4 by 1 by 1. A similar argument to that discussed in the general comments section about extending existing walls can be applied giving sequences

1, 1, 2, 3, 4, 6, 9, 13, 19, . . . $t_n = t_{n-1} + t_{n-3}$ $(n > 3)$,

and

1, 1, 1, 2, 3, 4, 5, 7, 10, 14, . . . $t_n = t_{n-1} + t_{n-4}$ $(n > 4)$.

Higher walls with 2 by 1 bricks

When building walls 3 units high with bricks 2 by 1 by 1 only even lengths are possible. Furthermore the argument about extending walls has to be modified as there are walls such as that in the diagram which cannot be seen as extensions of shorter walls of the same height.

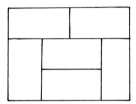

Fortunately it seems that only two such walls emerge for each new length and therefore recurrence relationships can be constructed. It has already been noted that there are three distinct walls 2 units high and 3 units long; turning these through $90°$ shows that there are equally three distinct walls which are 3 units high and 2 units long. Denoting these three A, B, C, walls 4 units long are AA, AB, AC, BA, BB, BC, CA, CB, CC + two new ones, the wall shown above and its horizontal mirror image.

So $t_4 = 3t_2 + 2$; by a similar argument $t_6 = 3t_4 + 2t_2 + 2$ and so on, giving the sequence 3, 11, 41, 153, 571, . . . which has the neater relationship $t_{2n} = 4t_{2n-2} - t_{2n-4}$.

How is this obtained or explained?

Walls 4 units high built with bricks 2 by 1 by 1 offer similar challenges to those 3 units high; all integral lengths are once again possible.

6   DIAGONAL

STATEMENT OF PROBLEM

On squared paper draw a rectangle 5 squares long and 3 squares wide. Draw a diagonal. Through how many squares does it pass?

Investigate other sizes of rectangle and analyse your results.

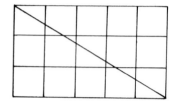

COMMENTARY

National Curriculum

The ability to draw rectangles (on squared paper) and the diagonal with reasonable accuracy is a fundamental requirement for tackling the investigation, which suggests at least level 3 and possibly level 4 (AT10). Underlying mathematical ideas involved in, and which should be strengthened by, the process of fully developing the investigation are **factors**, level 5 of AT5, and **enlargement**, level 6 of AT11.

Materials

Squared paper is essential for the drawings, and for recording and tabulation.

GENERAL COMMENTS

This is likely to be a well-known investigation to many teachers. It is included in this set of investigations because of its initial simplicity and appeal and because it has that classical quality of a good investigation which allows for the early development of a suitable conjecture which has then to be modified in the light of further enquiry.

It has proved excellent for group or class discussion; all pupils are rapidly involved in practical work and in finding a simple relationship between the number of squares through which the diagonal passes and the dimensions of the rectangle.

The investigation has appeared in print on a number of occasions, notably in Banwell *et al.*,[1] in which examples of childrens' work are included.

INITIAL APPROACHES

There should be no difficulty in arriving at an agreed answer to the initial question in the statement of the problem, particularly if each pupil draws the rectangle.

There is however a possibility that some pupils will interpret the word 'square' in 'through how many squares does it pass' to include not only the unit squares on the paper but also 2 by 2 squares and so on. It may then be argued that the statement of the problem should have been more precise. On the other hand, a certain amount of lack of precision in the statement of a problem may lead to useful preliminary discussion in deciding more clearly what should be investigated. Such discussion is helpful in focusing attention on the problem and cementing its main features in the mind. In the following development the unit square interpretation is used.

Once any initial difficulties have been resolved further data are then obtained by pupils drawing different sizes of rectangle. Invariably several examples of a given rectangle size will have been drawn thus ensuring eventual unanimity of observations after some rechecking and consideration of special cases.

One such special case concerns rectangles in which the diagonal passes through corners of squares other than at the corners of the rectangle itself. Various possibilities may be suggested, for example in the diagram:

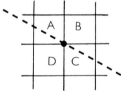

- count squares A and C only;
- count squares A, B, C, and D;
- count A and C and also the point as a square.

Any of these could be used but in the sections which follow the first suggestion counting A and C only is assumed.

ORGANIZING DATA

The question of how best to record the data systematically is, as always, an important stage. This may start with a list as shown in the table.

From such recording (or before) some specific generalizations begin to emerge such as $(n, 1) \to n$ and $(n,n) \to n$, though not necessarily in this algebraic form, and both of these are simple enough to explain.

A more efficient and helpful form of recording the data is discussed in a later section.

Dimensions		Squares on diagonal
(5, 3)	$\longrightarrow$	7
(7, 4)	$\longrightarrow$	10
(6, 6)	$\longrightarrow$	6
$\cdots$		

TOWARDS A CONJECTURE

A more generally applicable rule:

number of squares on diagonal = length + width − 1,

or $S = l + w - 1$, is soon put forward.

Not all rectangles, however, conform to this proposed rule, though they would if the third suggestion mentioned in the initial approaches section had been adopted in which points are counted also. Perhaps, therefore, anyone who proposes that method of counting has particular insight into the problem!

Attempts at explaining $S = l + w - 1$, where it applies, are to be encouraged; these often refer to 'moving across and down the rectangle', with the −1 being accounted for as 'not counting the corner twice'. A more precise explanation may use a diagram such as this one, using a stepping movement from corner to corner. The number of steps is $(l - 1)$ across and $(w - 1)$ down, with the number of squares being one more than the number of steps.

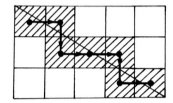

REFINING THE CONJECTURE

Non-conforming rectangles still remain to be considered. It is not spotted immediately that these all have the diagonal passing through one or more intermediate corners. Some may focus on the dimensions involved and suggest that the rule does not work when both dimensions are even, in which case it is a good strategy to suggest the checking of a counter example such as 6-by-3 so that the proposed hypothesis can be extended.

The pupil, however, who begins to employ ideas of similarity and refer to the 6-by-4 case as two 3-by-2 cases as shown in the diagram is well on the way to a full resolution of the problem.

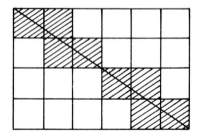

WELL-ORGANIZED TABULATION

A more concise and efficient way of recording the data is in the form of the following double-entry table. Such a table can be compiled on the blackboard as a class activity, and entries which do not conform to the $S = l + w - 1$ rule ringed. This type of table can provoke much fruitful discussion and productive insights.

						Length				
Width	1	2	3	4	5	6	7	8	9	10
1	1	2	3	4	5	6	7	8	9	10
2	2	2	4	4	6	6	8	8	10	10
3	3	4	3	6	7	6	9	10	9	12
4	4	4	6	4	8	8	10	8	12	12
5	5	6	7	8	5	10	11	12	13	10
6	6	6	6	8	10	6	12	12	12	14
7	7	8	9	10	11	12	7	14	15	16
8	8	8	10	8	12	12	14	8	16	16
9	9	10	9	12	13	12	15	16	9	18
10	10	10	12	12	10	14	16	16	18	10

Pupils arrive at this double-entry table method of recording themselves in due course, particularly if they have some experience of coordinates.

AN UNUSUAL INSIGHT

One quite extraordinary insight into this problem by an older student, who had studied mathematics to CSE level only, was based on an apparent attempt to generalize from the case of the square. Her method was to subdivide the rectangle into squares, at any stage marking off a square as large as possible as shown below.

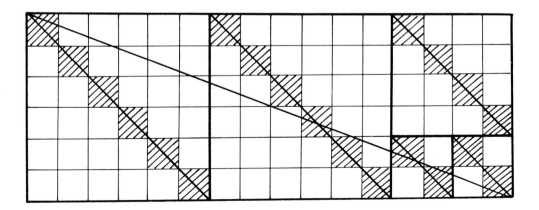

She then drew diagonals across each of the square subdivisions and asserted that the number of squares crossed by the diagonal of the rectangle is equal to the number crossed by all the diagonals of the square subdivisions:

Squares on the diagonal of the rectangle: 20.

Squares on the diagonals of the square subdivisions are 6, 6, 4, 2 and 2: $6+6+4+2+2=20$!

This striking assertion is fully justifiable by reference to the method for finding the highest common factor of two numbers by a process of dividing the smaller into the larger, then the remainder into the smaller and so on until the remainder is zero (Euclid's algorithm); the last non-zero remainder is the required highest common factor.

EXTENSIONS

Experience with this investigation suggests that extensions do not have much intrinsic appeal but this could be just a subjective viewpoint.

Counting all the squares through which the diagonal passes, not only the unit ones, is mentioned in the initial approaches section.

On triangled paper parallelograms can be outlined and the number of triangles through which either of the diagonals passes counted. The cases are different but both are closely related to the original rectangle problem.

An extension into three dimensions to find the number of cubes through which the diagonal of a cuboid passes is difficult to handle both visually and conceptually, though given that a general rule has been found for the rectangle the analogous rule for the cuboid follows readily.

REFERENCES

1 Banwell, C.S., Saunders, K.D. and Tahta, D.G. (1972) *Starting Points*, OUP, 6ff.

7 POSTAGE

STATEMENT OF PROBLEM

You have a good supply of 1p, 2p, 3p, 4p and 5p stamps. If you are allowed to put no more than two stamps on an envelope then all postage values up to 10p can be achieved. (The same stamp value can be repeated on an envelope.)

If your supply of stamps was different, say 1p, 2p, 4p, 5p and 6p, could you achieve all postage values up to 12p?

Can you get a higher maximum postage value by having a different set of five stamp values?

Vary the problem by having a different number of stamp values or allowing more stamps to be put on an envelope.

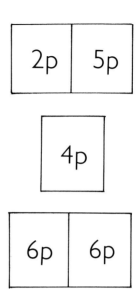

COMMENTARY

National Curriculum

The basic requirement is addition of whole numbers which is level 2 of AT3, but the ability to conduct an organized search is helpful which is level 4 (AT1). Thorough and careful search processes are a key feature as the investigation develops.

It also affords experience in AT5 when sequences of solutions begin to emerge.

Materials

Younger pupils will find a visual aid to represent stamps helpful, for example numbers on cards. Squared or lined paper is useful for recording.

GENERAL COMMENTS

The origins of this problem in its generalized form are unknown, but occasional articles appear on it in mathematical journals (for example Alter and Barnett,[1] 1980). Articles have also appeared in computing journals, no doubt because number-crunching search techniques would be very efficient for considerably larger numbers of stamp values than in the stated form of the problem. It is clear from these articles that many aspects of the problem remain unresolved. Nevertheless the problem has been found to intrigue a wide age-range of pupils and students.

Much can be achieved without recourse to using computers as the computation is simple. Some interesting patterns of results soon emerge and also some intriguing exceptions. Writing a simple program however, to expedite searches is a good challenge for the older pupil/student programmer.

INITIAL APPROACHES

A first strategy, frequently adopted, is to list all the integers (postage values) and the various ways in which they can be achieved using the given stamp values up to two at a time. It is then soon noticed that all values up to 12 can be achieved from the set $\{1, 2, 4, 5, 6\}$. Suspicion is aroused however by observing that many of the values can be obtained in two ways and 6 even in three; a higher maximum seems a definite possibility.

Some pupils then extend the integer list to include all ways in which each can be expressed as a pair of *any* two integers as shown in the table below, and this at first, can clarify the way forward, particularly with smaller sets of available values.

It can be seen from this table that all values up to 4 can be obtained from the sets $\{1, 2\}$ or $\{1, 3\}$, and that all values up to 6 can be obtained from $\{1, 2, 3\}$ (two other sets also do this), up to 7 from $\{1, 2, 5\}$ and up to 8 from $\{1, 3, 4\}$.

Postage	Stamp values			Postage	Stamp values				
1	1			6	6,	1+5,	2+4,	3+3	
2	2,	1+1		7	7,	1+6,	2+5,	3+4	
3	3,	1+2		8	8,	1+7,	2+6,	3+5,	4+4
4	4,	1+3,	2+2	9	9,	1+8,	2+7,	3+6,	4+5
5	5,	1+4,	2+3	10	10,	1+9,	2+8,	3+7,	4+6, 5+5

This covers all the sensible sets of three-stamp values $\{1, 4, 5\}$, for example not being sensible for obvious reasons, so a simpler problem than the one initially stated has been resolved.

With larger sets of available values, this method of observation becomes progressively more cumbersome and prone to error and an alternative strategy is looked for.

AN ALTERNATIVE STRATEGY

Another early strategy often used is to include only such stamps in the set as are absolutely necessary, in order to avoid the duplication of ways of obtaining particular postages. This strategy, it is argued, will make the set more 'efficient', that is the maximum postage value will be higher.

The value 1 has to be included in the set but 2 (given by $1 + 1$) does not seem necessary. Then 3 is included and so 4 is not required. For a set of three stamps it is then supposed that $\{1, 3, 5\}$ will give the best result. Unfortunately if the line of investigation in the previous section has been followed, both $\{1, 2, 5\}$ and $\{1, 3, 4\}$ are known to be more efficient, so the plot thickens and the strategy must be adjusted in some manner.

SETS OF FOUR STAMPS

Moving on to sets of four stamps is natural at this stage. A ploy often pursued is to add one stamp to the best of the sets of three $\{1, 3, 4\}$. Anyone of 5, 6, 7, 8 or 9 can be sensibly added with three of the resulting 4-stamp sets raising the maximum postage value to 10. The advance, from the maximum of 8 for sets of three, to 10 for sets of four is often regarded as disappointing and this gives rise to a period of mulling and random experiment. The table in the previous section is consulted again and extended beyond 10; this often turns out to be one of the happy 'eureka' experiences in mathematical investigation. Suddenly it is spotted that $\{1, 3, 5, 6\}$ will give a maximum of 12.

Amongst observations made about this surprising result are that:

- the set has been formed by the addition of a number to one of the least efficient sensible sets of three stamps;
- the addition of 6 to the set $\{1, 3, 5\}$ has doubled to 12 the maximum obtained by that set;
- the maximums form a sequence as shown in the following table, which suggests a possible maximum for sets of five stamps of 16, and this is significantly better than for the sets mentioned in the initial statement.

Number of stamps	Most efficient set	Maximum postage value
2	$\{1, 2\}$ or $\{1, 3\}$	4
3	$\{1, 3, 4\}$	8
4	$\{1, 3, 5, 6\}$	12
5	?	16?

Further exploration of other sets of four stamps reveals:

- the interesting absence of 11 as a possible maximum;
- several other sets giving a maximum of 10, for example $\{1, 2, 5, 8\}$.

SETS OF FIVE STAMPS

The search for the postulated maximum of 16 for a 5-stamp set can be developed from results already established for 4-stamp sets. Using the strategy of adding another stamp value to the best 4-stamp set of $\{1, 3, 5, 6\}$ gives a maximum of 14 by the addition of 7, 8 or 13. Once again the increase in the maximum seems disappointing.

Following up another of the observations of the previous section suggests adding a stamp to one of the least efficient 4-stamp sets. Of those previously considered $\{1, 3, 4, 7\}$ has a maximum of only 8, but $\{1, 3, 4, 7, 8\}$ and $\{1, 3, 4, 7, 9\}$ give maximums of only 12 and 14 respectively with no further possible additions being sensible. Once again a way forward is not obvious and while a systematic search of all possible sensible 5-stamp sets may be suggested it is then found to be quite extensive and arduous.

Trying to extend the sequence of the most efficient sets table in the previous section eventually results in consideration of $\{1, 3, 5, 7, 8\}$ which does have the postulated maximum of 16. This discovery always appears as a significant breakthrough. A pattern is now much more obvious in the resolution of the problem and the extension to a 6-stamp set of $\{1, 3, 5, 7, 9, 10\}$ with a maximum of 20 is quickly forthcoming.

ANOTHER STRATEGY FOR 5-STAMP SETS

A strategy sometimes adopted if attention remains focused on 5-stamp sets from the start, is to spread out, gradually, the range of the set, often concentrating at first on the larger stamp values.

For example $\{1, 2, 4, 5, 6\}$ with a maximum of 12
is spread to $\{1, 2, 4, 6, 7\}$ with a maximum of 14
and then to $\{1, 2, 4, 7, 8\}$ with a maximum of 12 (fails at 13)
and then to $\{1, 2, 4, 7, 9\}$ also with a maximum of 14.

Then a $\{1, 3, \ldots\}$ start is considered and the optimum set $\{1, 3, 5, 7, 8\}$ eventually found by trial and error rather than by the developing pattern of solutions of the previous section. A hypothesis frequently arising out of this type of search is that in optimum cases the largest two values in the set must be consecutive. Though this often is the case it will be dismissed subsequently and is less of a hindrance to exploration than the 'doubling rule' trap mentioned in the next section.

OTHER DEVELOPMENTS

At some stage in the investigation the *minimum* postage which can be achieved rather than the maximum is considered and then thought to be too simple to be worthy of much attention; as in the statement of the problem $\{1, 2, 3, 4, 5\}$ gives all values up to 10. Clearly, it is argued, this must be the most inefficient sensible set. The meaning of 'sensible' however needs some clarification. (Missing 1 out or starting 1, 4 are neither seen as sensible and so on.)

Another observation is that both the most efficient and most inefficient sets have as their maximum double the largest number in the set, for example

$\{1, 2, 3, 4\}$ gives all postage values up to 8,

and

$\{1, 3, 5, 6\}$ gives all postage values up to 12.

This idea seems quite plausible but has pitfalls. Assuming it always to be the case can lead to thinking that the problem has been neatly resolved in general. Looking back a little will show however that sets between the two proposed extremes do not necessarily follow such a rule, with examples such as $\{1, 2, 4, 6\}$ and $\{1, 3, 4, 7\}$ giving only postage values up to 8. If these do not follow the doubling rule why should the most efficient set necessarily follow such a rule? The problem has more surprises in store.

SETS OF SIX STAMPS

The more adventurous investigators move on to sets of six stamps. The analysis so far indicates a possible most efficient set of $\{1, 3, 5, 7, 9, 10\}$ with all postage values up to 20. Other approaches, for example exploring stamp sets that start $\{1, 2, \ldots\}$ eventually generate another solution $\{1, 2, 5, 8, 9, 10\}$ which still follows the potentially misleading doubling rule. Nevertheless the existence of two optimum solutions provokes the search for more and they do indeed turn up and none of them follow the doubling rule, for example $\{1, 3, 5, 6, 13, 14\}$. The shackles are undone.

Not only is the doubling rule undone but so is the pattern of the table in the section on sets of four stamps. The optimum of $\{1, 2, 5, 8, 9, 10\}$ for six stamps can be extended to $\{1, 2, 5, 8, 11, 12, 13\}$ for seven stamps with all stamp values obtainable up to 26, better than 24 from $\{1, 3, 5, 7, 9, 11, 12\}$. Other patterns of solutions analogous to that in the table can be formulated which do not necessarily always represent optimum solutions.

EXTENSIONS

If no *repeated* stamp values are permitted on an envelope then the problem clearly has some affinity with the traditional set of weights used in the days of imperial measure when household scales were equipped with a set of 1-, 2-, 4-, 8-, 16- and 32-ounce weights. Thus a 3-stamp set of $\{1, 2, 4\}$ gives a maximum of 6 with up to *two* stamps on an envelope; a 4-stamp set of $\{1, 2, 4, 8\}$ gives a maximum of 14 if up to *three* stamps are allowed on an envelope and 15 for up to *four* on an envelope. However if up to two stamps only are allowed from a 4-stamp set then adjustments have to be made to the 'weights' set.

Another variation proposed is to look for sets of stamps which give the longest consecutive run of postage values. For example using up to two stamps from $\{2, 3\}$ with repeats allowed gives all postage values from 2 to 6, a run of 5 values. This is better than either $\{1, 2\}$ or $\{1, 3\}$.

Three or more stamps on an envelope

This is the most promising extension, having already been mentioned briefly above and also in the statement of the problem. With repeats being once again allowed the maximum postage achieved using three, four, five or more stamps from a set of two can result in an easily determined pattern of solutions which older pupils and students possibly can generalize algebraically.

Number of stamps used	Stamp set	Maximum postage value
up to 2	$\{1, 3, 4\}$	8
up to 3	$\{1, 4, 5\}$	15
up to 4	$\{1, 5, 6\}$	24
up to 5	$\{1, 6, 7\}$	35

With a 3-stamp set the optimum solution $\{1, 3, 4\}$ for up to two stamps at a time forms the start of a pattern of solutions which plausibly looks optimal in general; this turns out on further investigation to be an incorrect supposition. Other patterns of solutions can be used to find higher maximums such as 26 for up to four stamps from $\{1, 5, 8\}$. Once again the problem had a surprise in store and this is an essential part of its fascination.

REFERENCES

1 Alter, R. and Barnett, J.A. (1980) A postage stamp problem, *American Mathematical Monthly*, March, 206–10.

8 POCKET

STATEMENT OF PROBLEM

On this unusual billiard table there are just four pockets, one at each corner; it measures 4 units by 6 units. The diagram shows the path of a ball fired from A at 45° which, after rebounding from three sides, drops into the pocket at D.

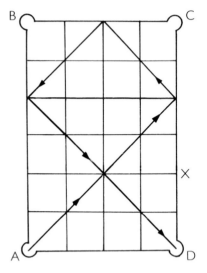

On squared paper draw another such billiard table 5 units by 6 units and show on it the path of a ball fired from A at 45°. Down which pocket does it drop? After how many rebounds?

Investigate what happens with other sizes of billiard table, or what happens if the firing angle is changed, for example so as to bounce first at X, in the diagram.

COMMENTARY

National Curriculum

The ability to draw accurate diagrams on squared paper suggests a minimum of level 3 for this investigation. Though degrees are not mentioned until level 5 (AT8), the reference to the eight points of the compass in level 3 of AT11 is clearly sufficient.

Nevertheless pupils at level 2 can enjoy exploring the problem if one of the computer simulations is used.

In the full development described here the investigation uses and gives good experience of **factors** (level 5, AT5), **ratio** (level 6, AT2), **symmetry** and **reflection** (level 3, AT11), and **enlargement** (level 6, AT11).

Material

Squared paper is essential for the drawings, and useful for recording and tabulation.

GENERAL COMMENTS

The problem has considerable basic motivation because of its analogy with billiards and snooker. It is likely to be as well-known as the investigation DIAGONAL (6) with which it has a structural connection.

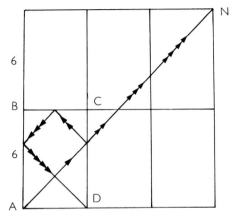

This connection can be explained with the help of this diagram. The 4-by-6 rectangle has been tessellated until a square is produced. This occurs when a square of side 12 units is obtained, 12 being the lowest common multiple of 4 and 6, to be denoted LCM (4, 6). The number of rebounds on the path of the ball from A to D is one less than the number of sections of the path. The path from A to D is 'mirrored' by the diagonal AN of the square and each of the four sections of the path corresponds to one of the tessellated rectangles through which the diagonal passes.

In DIAGONAL (6), squares are tessellated; what we have here is a linear transformation (one-way stretch) of that problem which leaves its nature unaltered.

Using the result of the diagonal problem 'length + width − 1' for the number of parts of the diagonal and generalizing for an m-by-n billiard table, the side of the square is $\mathrm{LCM}(m, n)$ and the numbers of rectangles along each side are

$$\frac{\mathrm{LCM}(m, n)}{m}, \quad \frac{\mathrm{LCM}(m, n)}{n}.$$

Hence the number of rebounds is

$$\frac{\mathrm{LCM}(m, n)}{m} + \frac{\mathrm{LCM}(m, n)}{n} - 2.$$

The two numbers $\dfrac{\mathrm{LCM}(m, n)}{m}, \dfrac{\mathrm{LCM}(m, n)}{n}$

have by definition no common factor and hence the use of 'length + width − 1' is generally justified.

An alternative form of this result in terms of the highest common factor of m and n, denoted HCF(m, n), is

$$\frac{(m+n)}{\text{HCF}(m, n)} - 2.$$

There are several versions of a computer program which simulate this problem. A good feature of these programs is that as the path of the ball is drawn it can be halted temporarily in mid-course. This is particularly useful when using the program for a group or class discussion.

INITIAL APPROACHES

Some initial experimentation in producing correct drawings on squared paper may be necessary. For consistency of results concerning the pocket a convention is necessary about which dimension of the billiard table is the width and which the length and about how the pockets are labelled. In this commentary the width is down and the length across the page and the labelling of pockets is clockwise starting from bottom-left, as shown in the diagram in the statement of the problem.

Then, group work is appropriate to draw examples systematically in order to produce enough data to show some pattern in the relationship between table dimensions, the number of rebounds and the pocket down which the ball drops.

Pupils should be encouraged to develop their own methods of organizing the data. One ingenious method which has been used for making a composite drawing for a billiard table of given width but varying lengths is shown in the following diagram, in which the paths of the ball on all such tables can be followed easily.

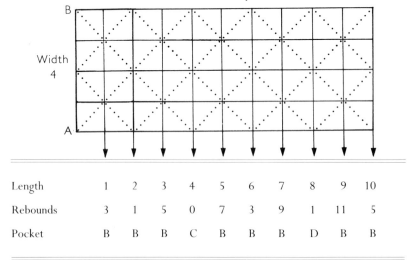

Length	1	2	3	4	5	6	7	8	9	10
Rebounds	3	1	5	0	7	3	9	1	11	5
Pocket	B	B	B	C	B	B	B	D	B	B

Of course, this figure can be used for tables of width less than 4 and so a composite diagram for all possible tables up to 10 by 10 say, is quite feasible.

FIRST CONJECTURES

Whether or not such a composite method of obtaining data is used, suggestions about the relationships are soon forthcoming. For example, in the diagram in the previous section it can be seen that, for tables of width 4, the number of rebounds is 2 more than the length when this is odd. The pocket is B except when the length is a multiple of the width, in which case an extension of the table can be visualized to show that the pocket is C if the length is an odd multiple of the width and D if it is an even multiple. Some more general possible observations are mentioned at the end of the next section and some of these may well be put forward without necessarily involving the detailed data tabulation of that section.

FURTHER CONJECTURES

The similarity of the path on a 3-by-2 table to that on the 6-by-4 table (as in the statement of the problem) may be noticed and generalized to any integral enlargement of a 3-by-2 table. Extending this idea to a generalization for all sizes of table, this suggests division of the width and the length by their highest common factor. It could be helpful to produce full tabulation of results for tables up to 10-by-10, and this is not too great a task particularly if a composite path diagram mentioned in the initial approaches section is used.

| | Number of rebounds | | | | | | | | | | | Pocket | | | | | | | | | |
Length	1	2	3	4	5	6	7	8	9	10		1	2	3	4	5	6	7	8	9	10
Width																					
1	0	1	2	3	4	5	6	7	8	9		C	D	C	D	C	D	C	D	C	D
2	1	0	3	1	5	2	7	3	9	4		B	C	B	D	B	C	B	D	B	C
3	2	3	0	5	6	1	8	9	2	11		C	D	C	D	C	D	C	D	C	D
4	3	1	5	0	7	3	9	1	11	5		B	B	B	C	B	B	B	D	B	B
5	4	5	6	7	0	9	10	11	12	1		C	D	C	D	C	D	C	D	C	D
6	5	2	1	3	9	0	11	5	3	6		B	C	B	D	B	C	B	D	B	C
7	6	7	8	9	10	11	0	13	14	15		C	D	C	D	C	D	C	D	C	D
8	7	3	9	1	11	5	13	0	15	7		B	B	B	B	B	B	B	C	B	B
9	8	9	2	11	12	3	14	15	0	17		C	D	C	D	C	D	C	D	C	D
10	9	4	11	5	13	6	15	7	17	0		B	C	B	D	B	C	B	D	B	C

Various conjectures are possible considering the right-hand side of this table which can then be justified by reference to the path diagram:

- if both width and length are equal or odd then the pocket is C;
- if the width is odd and the length is even then the pocket is D;
- if the width is even and the length is odd then the pocket is B; this is an inverse form of the previous conjecture, for example on a 3-by-4 table the pocket is D and on a 4-by-3 table it is B;
- if both length and width are even then reduction by an even factor makes it equivalent to one of the first three cases.

If instead one focuses only on rebound entries where the length and width are co-prime, that is have a highest common factor of 1, then the result 'length + width − 2' is observed, and all other sizes of table reduce to one of this type by scaling down by the HCF.

An explanation of the 'length + width − 2' result has already been outlined in the general comments section but other ideas, partially or completely formulated, may be suggested and are to be encouraged.

Other variables to investigate

In addition to the pocket and rebound variables mentioned in the statement of the problem there are other features to be investigated such as the length of path and the number of lattice points (corners of squares) over which the ball passes. The former of these is clearly in general $\sqrt{2} \times \text{LCM}(m, n)$ (see diagram in the general comments section). Younger pupils could measure it in 'square diagonals'.

Changing the firing angle

Varying this is an obvious extension, for example halving the gradient (though not the angle) of the first path, to fire at X in the diagram in the statement of the problem. Results in this case are strongly related to those already obtained; a 6-by-4 table has effectively become a 12-by-4 table. It is best to aim initially at a corner of a square along the opposite edge of the table. Even so, with some tables subsequent rebounds on the top and bottom sides of the table may not be at corners of squares and some care is needed to produce an accurate drawing. If this is done on squared paper then squares outside the table can be used to determine the correct direction for a rebound.

Clearly *some* initial angles of fire will result in an interminable path.

Firing from other positions

In some cases this can result in a continually repeated path, for example when firing at $45°$ from the corner of the square to the right of A in the diagram in the statement of the problem. When and why such a path occurs could be investigated.

Triangular tables

A more challenging extension is to consider equilateral triangular tables subdivided into smaller such triangles; using triangled paper enables correct lines for the rebounds to be determined. Size of table, in this case, is not a variable, only the angle of fire.

9   CHEESE

STATEMENT OF PROBLEM

Consider slicing a flat square slab of
cheese into nine equal square pieces.
It would appear to need four straight
cuts as shown in the first diagram in
which the dotted lines are the cuts.

If you are allowed to rearrange the
pieces between each straight cut,
could you complete the task with less
than four cuts?

For example, if the first cut is along
XX then the next cut could be as
shown in the second diagram.

Extend the problem by considering other numbers of pieces and different initial slab shapes of cheese; for
example cutting a 5-by-3 slab into 15
equal square pieces. In each case try
to make as few cuts as possible.

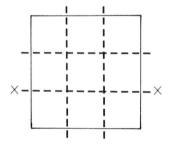

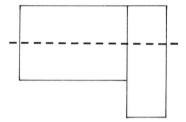

COMMENTARY

National Curriculum

Level 2 (AT11) refers to turning through right angles and level 3 (AT10) to sorting 2-D
shapes; these are sufficient to start this investigation.

Materials

Squared paper is helpful for recording each stage in the slicing. If required, cut-out figures
can be used to consider the possible rearrangements of pieces between each cut.

Centicubes could also be used to model the problem.

GENERAL COMMENTS

Taking a simpler problem first, to cut the square of cheese into four equal square pieces clearly needs just two cuts. The nine-piece problem with which the statement of the investigation opens may be familiar. It has appeared from time to time in the 'brain teaser' category with an elegant solution which few solvers would discover even after much cogitation. It is argued that the number of cuts has to be four because the centre square requires one distinct cut for each of its sides regardless of any rearrangement of pieces which might take place between cuts. A similar argument can apply to other slabs; for example in a 2-by-3 slab the two centre squares each need a minimum of three cuts.

It was the discovery that to cut a square of cheese into sixteen equal square pieces could also be achieved in only four cuts which prompted further investigation; the problem suddenly became more interesting!

INITIAL APPROACHES

Experimental work should include various rectangular slabs as well as square slabs of cheese and these should be considered systematically, perhaps sticking to the one-dimensional case first, 1-by-2, 1-by-3, 1-by-4 and so on; this is discussed in more detail in a subsequent section.

Young pupils can cut the slab out of squared paper, slice it, then consider possible arrangements physically before making the second cut.

Note that if centicubes are used to model the problem it is the size of slab which varies, with the small square-piece size remaining constant. However this, though different from the way in which the problem is initially stated, is effectively the same.

TABULATING DATA

When sufficient data has been accumulated there is the usual problem of how best to organize it. Pupils will not immediately resort to a double-entry table without previous experience; the commentary on DIAGONAL (6) deals with this dilemma in more detail. Tabulating data in such a form is a very helpful step to take because it shows up patterns and also acts as a check on consistency of observations.

	Length							
Width	1	2	3	4	5	6	7	8
1	0	1	2	2	3	3	3	3
2	1	2	3	3	4	4	4	4
3	2	3	4	4	5	5	5	5
4	2	3	4	4	5	5	5	5
5	3	4	5	5	6	6	6	6
6	3	4	5	5	6	6	6	6
7	3	4	5	5	6	6	6	6
8	3	4	5	5	6	6	6	6

The table above shows the minimum number of cuts required for initial slab proportions of up to 8-by-8. Much practical work is involved in obtaining this data and many pupils will only achieve a subset of it, either up to 4-by-4, or just the first row (or column) for the one-dimensional case.

THE ONE-DIMENSIONAL CASE

Developing the one-dimensional case further than shown in the table is a useful approach as eventually it shows how powers of 2 have a bearing on the results; all the sizes, from 1-by-9 to 1-by-16, require a minimum of four cuts but 1-by-17 and so on require five. This observation can be explained by reference to a systematic process of cutting and rearranging after each cut.

In all cases the position of the first cut will be found to be crucial for the minimum to be achieved. Where this is, and why, should be considered and explained.

ANALYSING THE DATA TABLE

To achieve the results for the whole table could be a small-group activity, sharing the work involved and discussing any results which do not appear to be consistent with the increasingly evident pattern in the table as it is compiled. An important principle to establish is that a particular size of slab cannot have a minimum number of cuts which is less than that for a slab size which it can contain. This may seem a trifle obvious to some but it has been found not always to be so.

Once the full table of results is agreed, it is possible to speculate on how the pattern would grow in a larger table and conduct further experimental cuttings to confirm or reject hypothesized results.

Another aspect of the table, which is often noted, is that more pieces does not necessarily mean more cuts; for example slicing a 3-by-5 slab into 15 pieces takes one more cut than a 4-by-4 slab sliced into 16 pieces. This observation has been extended to a conjecture which asserts that for a given number of pieces the number of cuts is independent of the slab shape. Both 4-by-6 and 3-by-8 slabs need five cuts and from an extended table it will be found that the same is true for 2-by-12 and 1-by-24 slabs. This seems a plausible hypothesis but unfortunately it has exceptions. Can it be adjusted to take the exceptions into account?

INDEPENDENCE OF DIMENSIONS

A more complex observation which some pupils may suggest in a carefully conducted discussion is that the table is an addition table using the second row and column as 'master' row and column, for example 5 cuts for the slab size 4-by-6 is found by adding 2 (for 1-by-4) and 3 (for 6-by-1). This is because the two dimensions can in effect be treated separately and explanations of this by reference to the physical process of cutting are possible.

Eventually such an explanation could lead to an algebraic formulation (for those interested in such aspects) for the minimum number of cuts for the m by n rectangle as:

$$[\log_2(m-a)+1]+[\log_2(n-a)+1]$$

where $m>1$, $n>1$, $0<a<1$, and where $[x]$ means the integral part of x.

EXTENSIONS

Triangular pieces

Cutting the cheese into right-angled isosceles triangles (having angles $90°$, $45°$, $45°$) is perhaps the most obvious extension. The simplest answer is that it always takes one more cut after first cutting the slab into squares, akin to slicing a pile of square sandwiches diagonally.

However there are other aspects of this development; a 2-by-2 slab with two diagonal cuts gives four of the required triangles each of area one square as shown in the diagram. A 2-by-4 slab can be cut into eight such triangles with three cuts and a 2-by-3 slab can be cut into four such triangular pieces and two unit square pieces also with three cuts.

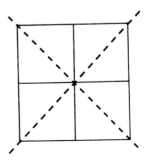

Can any slab size always be sliced into a combination of such triangles and unit squares with the same number of cuts as for squares only?

Three dimensions

The three-dimensional problem of cutting a cuboid into cubes is worth considering but the tabulation of results needs care. If the idea of independence of dimensions has been reached then predictions and theories for three dimensions are fairly straightforward. A discussion of the practical problems and eventual theorization can be found in Krause[1].

A link with deltominoes

There is a family of shapes which can be made by fitting equilateral triangles together edge to edge including of course larger equilateral triangles; see the investigation DELTOMINOES (3). The problem of slicing these into equilateral triangles also has a threefold aspect to it and is quite challenging; it has analogies with both the original and the cuboid problems.

REFERENCES

1 Krause, E.F. (1983) The caterer's problem, *Mathematics Teacher* (NCTM), March.

10   RANDOM DOTS

STATEMENT OF PROBLEM

Here are two random arrangements of four points.

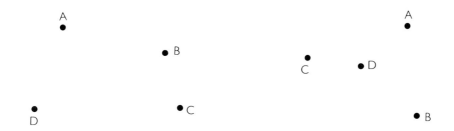

If each of these is joined, A to B to C to D to A (ABCD for short) then two different quadrilaterals are formed. How do they differ?

What happens if, instead, you join up the points in the order A*CBD* in each of the two arrangements?

If five points, no three of which are in a straight line, are randomly placed on a flat piece of paper, is it always possible to join up four of the points to form a convex quadrilateral?

What arrangement of the five points gives the greatest number of convex quadrilaterals? What arrangement gives the least?

COMMENTARY

National Curriculum

There is no explicit reference in the Attainment Targets to either basic drawing skills such as joining two points with a straight line, or to the idea of **convexity**; both are implicit in level 4 of AT10 which refers to the language of angle and to constructing figures. However level 3 (sorting 2-D shapes) would seem to be the minimum for this investigation.

To go beyond five points (not suggested in the statement of the problem) needs combinatorial skills of level 6 and upwards.

Materials

Plain paper is best for the drawings.

Some simple method of creating a pin-jointed quadrilateral, for example strips of card or folded paper linked with paper fasteners, is useful for discussing different shapes of quadrilateral.

GENERAL COMMENTS

The two arrangements or configurations of four points in the statement of the problem are fundamentally distinct because the first gives a convex quadrilateral (all interior angles being less than 180°), whereas the second does not. Quite young pupils will readily classify quadrilaterals into two types, convex and not convex.

Another fundamental question is that of the number of ways in which four points can be joined up in succession. The convex quadrilateral in the first configuration can be drawn in eight different ways, two ways starting at each of the four vertices. There are, of course, 24 permutations of A, B, C, D and eight of these correspond to the eight ways of drawing this convex quadrilateral. The remaining 16 ways correspond to two *crossed* quadrilaterals as shown in this diagram.

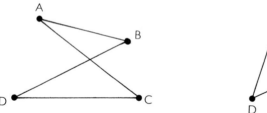

Some pupils may object that these are not quadrilaterals at all! This is because they may be seen as offending the conventional, simple closed planar curve form of a quadrilateral. Emphasizing the vertices may help here.

For the second configuration in the statment of the problem, three distinct quadrilaterals can again be drawn, this time they are all non-convex ('**concave**' seems a suitable word?).

Consequently we have, for any configuration of four points, either

- one convex and two crossed quadrilaterals, or
- three concave quadrilaterals.

INITIAL APPROACHES

An initial group or class discussion on the idea of convexity may well be appropriate; initially this could focus on curved figures. A formal definition of convexity is not necessary at this stage; pupils can understand the concept without recourse to such a definition.

Free experimentation with the five-point problem is useful for gaining familiarity with the ideas involved. Configurations should be recorded with the possible convex quadrilaterals drawn in different colours.

Some consideration should be given to the question of how many different quadrilaterals can be drawn in this five-point case. One way is to say that there are five ways of choosing four points (i.e. five ways of discarding one) and each of these five gives rise to three quadrilaterals, so that we have fifteen altogether.

CLASSIFYING FIVE-POINT CONFIGURATIONS

An initial attempt at classification might be along the lines of:

- convex pentagon of points;
- convex quadrilateral of points with the fifth point inside the quadrilateral;
- triangle with two points inside.

Whether or not this is an exhaustive classification of the possibilities for five points needs some discussion. Does it matter where the internal points are, inside the quadrilateral or triangle?

An exhaustive method of classification put forward by older pupils is achieved by considering the addition of one point to either of the two original four-point configurations in the statement of the problem. First of all, the six possible lines joining two points in the four-point configurations are drawn as shown below. The fifth point is then placed in turn in each of the regions of the plane so produced.

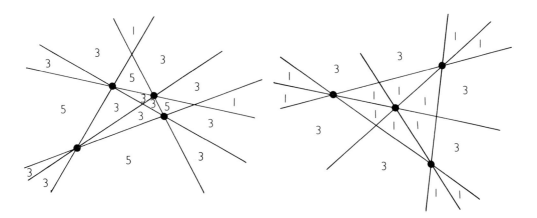

The significance of these six lines is that in moving the fifth point from one region into an adjacent one changes some possible convex quadrilaterals into concave ones and vice versa. It also changes some concave quadrilaterals into crossed ones and vice versa. This seems to suggest an ordering of the three types of quadrilateral (convex, concave, crossed) which can be illustrated neatly by the jointed framework of four strips referred to above in the section on Materials; a convex quadrilateral has to become concave before it can become crossed!

Note that in the diagram all possible intersections of these six lines have been included. Some of these could be omitted accidentally in the first diagram, if either pair of opposite sides of the quadrilateral were nearly parallel.

The numbers in the regions correspond to the number of convex quadrilaterals which can be drawn with the fifth point placed in that region. Obviously some 'symmetry' is present.

From these diagrams it would appear that we now have a clear demonstration that precisely three distinct configurations are identifiable corresponding to the 1, 3 and 5 possible convex quadrilaterals, thus confirming the classification already tentatively made.

Configuration	Convex	Crossed	Concave	Total
Convex pentagon	5	10	0	15
Convex quadrilateral with 1 internal point	3	6	6	15
Triangle with 2 internal points	1	2	12	15

SIX-POINT CONFIGURATIONS

The six-point problem can now be investigated by following some of the approaches suggested for the five-point case. To complete a full analysis is a considerable challenge, possibly as a group activity.

Extending the approach of the previous section requires separate consideration of each of the three different configurations of five points and diagrams such as the one below, based on the five-point configuration of a triangle with two internal points, can be produced. Again it is important to make sure that all possible regions are included; each line joining two of the five points should intersect all three sides of the triangle formed by joining the other three points. Getting the correct number for each region needs a good deal of care; all possible sets of three points (there are ten from the five given points), with which the sixth point is to be linked to form a quadrilateral, should be listed and each quadrilateral checked for convexity. Once some numbers are agreed a pattern emerges which can act as a check for consistency, for example moving from one region to an adjacent one changes the number by either one or

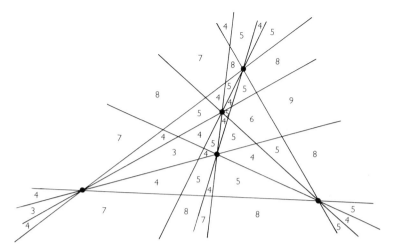

three. A change of three occurs only when crossing one of the lines forming the outer triangle of three points. This observed pattern may need much discussion before it is explained satisfactorily.

For the various six-point configurations considered in the diagram, there are any number from three to nine convex quadrilaterals possible; the two other five-point configurations also need to be analysed similarly. In general it would appear that the minimum number of convex quadrilaterals possible is three and the maximum 15. A curious result is that no configurations appear to give rise to either 13 or 14 convex quadrilaterals.

CLASSIFYING SIX-POINT CONFIGURATIONS

An attempt could be made to distinguish all the distinct six-point configurations as already done with five points, starting with the 'convex hexagonal ring' which gives the maximum of 15 convex quadrilaterals. This particular type of configuration is one that clearly can be generalized for n points giving $^nC_4$ convex quadrilaterals, where $^nC_4$ is the number of different combinations of 4 points chosen from n points.

The next configuration to consider would seem to be a convex pentagon with one internal point. This configuration, however, can give rise to 10, 11 or 12 convex quadrilaterals. (This result will have emerged from the analysis in the previous section if all three five-point configurations have been analysed.) So, for this particular configuration three distinct sub-configurations need to be identified and described.

Clearly this classification exercise is not straightforward. A convex quadrilateral with two points inside can given rise to 7, 8 or 9 convex quadrilaterals, which can be seen fairly easily from the diagram in the previous section.

Finally, a triangle with three internal points can give 3, 4, 5 or 6 convex quadrilaterals.

Is it possible to describe unique configurations corresponding to each of the possible numbers of convex quadrilaterals?

A THEORY ABOUT MINIMUM NUMBERS

In the previous section the maximum possible number of convex quadrilaterals for a given number of points is generalized. The minimum case appeared as a problem in the 1969 International Mathematical Olympiad and is detailed by Watson[1].

Two members of the British team, D.J. Aldous and N.S. Wedd (sixth-form pupils) suggested the generalized result that there are at least $^nC_5/(n-4)$ convex quadrilaterals (CQ) for n points. The argument ran something like this.

> For each set of 5 points there is at least one CQ. For n points there are $^nC_5$ sets of 5 points; however, two different 5-point sets may give rise to the same CQ. Each such quadrilateral could come from at most $(n-4)$ sets of 5 points. Hence the minimum number of CQs is at least $^nC_5/(n-4)$.

For $n=6$ this formula gives 3 as the minimum which accords with the result of the section on six-point configurations.

For $n=7$ the formula gives 7.

EXTENSIONS

To treat the seven-point case in the same manner as the two previous cases seems a daunting task; some other approach is generally needed. However the minimum case has been looked at as an extension of the corresponding minimum six-point configuration; the least number of convex quadrilaterals is conjectured as 9, which is a higher figure than that predicted by the theory of the previous section.

Pentagons

An extension to convex and other types of pentagon is another possibility. The first point to note is that there is not just one type of concave pentagon. Such a pentagon can have either one or two reflex angles, and if two, then these can be either adjacent or not. Likewise, crossed pentagons, one of which is the well-known pentagram, are not all of the same type. These questions form an investigation in its own right and could be extended to consider the different possible types of hexagon.

REFERENCES

1 Watson, F.R. (1970) A problem on convex quadrilaterals, *Mathematical Spectrum*, **2** (2).

11   CUSTARD PIES

STATEMENT OF PROBLEM

A number of people are standing in a room (with no carpet for ease of cleaning) each armed with one custard pie. No two of them are the same distance apart as any other pair.

At a given signal each person throws their pie with deadly accuracy at the nearest person. Assume that if two people throw pies at each other then their pies do *not* collide.

Where would it be best to stand in order not to be hit?

Analyse the results of various fights, varying the number of participants and their positions within the room.

COMMENTARY

National Curriculum

Level 5 of AT11 refers to using **networks** to solve problems which is clearly relevant to this investigation. Drawing fight configurations and distinguishing between them involves simple ideas of network or graph theory and the concept of **topological equivalence** (see comments on ATs 10 and 11 in Chapter V, National Curriculum). Using these ideas, evidence would suggest that pupils at level 3 can explore successfully some of the early stages of the investigation with 'fights' involving up to four persons.

Levels 4 and 5 of AT10 include the construction and angle properties of triangles. The triangle geometry involved later in the development of the investigation concerns inequality, for example that the longest side of a triangle is opposite the largest angle, a result which is presumably also level 4 or 5.

Materials

Physical enactment of possible fights is perhaps unwise; pointing at the nearest person is safer. Recording of fight networks is best on plain paper with squared paper being used for tabulating when that arises.

GENERAL COMMENTS

If the aggressive nature of the 'model' used to motivate the problem is disliked then it can easily be replaced. For example, a similarly structured problem appeared in the *American Mathematical Monthly*[1]:

> *n* birds land at random in a field. Each bird watches its nearest neighbour. What is the expected number of unwatched birds?

Another basic, but often neglected, theorem in Euclidean geometry, akin to the result already mentioned in the previous section, is the 'triangle inequality', that the sum of the lengths of any two sides of a triangle is greater than the length of the third side. Use of this result by older pupils and students can formally explain some observed features of these pie fights which otherwise prove difficult to account for convincingly.

INITIAL APPROACHES

The simple case of just two people, who clearly must throw at each other, rapidly makes way for the three-person case. The diagram shows a popular way of illustrating a possible custard-pie fight between three persons, with the paths of the pies shown by arrows; where two persons throw their pies at each other the arrows are curved to avoid each other to conform to the rule in the statement of the problem that the pies do not collide. (This rule could, of course, be changed.) The diagram can be called a fight-network or fight-graph.

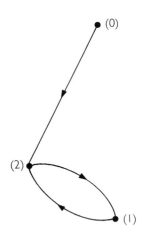

It is often speculated whether a three-person fight could occur in which all three persons receive hits. Experimenting with random positions of three points seems to suggest that this cannot occur; pupils formulate a variety of sensible explanations why it cannot. However they may need to draw a variety of different relative positions for the three persons before coming to the conclusion that all three-person fights can be classified as just one *type* in which one person always receives two hits, another one and the third none. A typical notation devised for this fight is $(2, 1, 0)$.

Some pupils want to distinguish between the six different ways this type of fight can occur between three specific persons, corresponding to the six different ways of labelling the dots in the above diagram, A, B and C.

FOUR-PERSON FIGHTS

With four persons the number of different types of fight increases significantly and a method of checking for all cases is soon seen to be necessary. One way in which this is done is to extend the notation already invented for three-person fights. Four different types of four-person fight graphs are found by experiment and using this notation these can be represented as

$$(3, 1, 0, 0) \quad (2, 2, 0, 0) \quad (2, 1, 1, 0) \quad (1, 1, 1, 1)$$

where an arbitrary convention has been adopted to arrange the numbers in descending order.

Each of these sets of four numbers adds up to 4 (for obvious reasons) and this suggests one possible way by which all types of four-person fight-graphs might be found systematically. How many ways can 4 be split up into four numbers from the set $\{0, 1, 2, 3, 4\}$ with repeats being allowed? The four 'quadruples' already listed are four of five such ways with $(4, 0, 0, 0)$ missing from the list; this however is as soon rejected as the description of an impossible fight and it would appear that all types of four-person fight have indeed been found.

GENERAL OBSERVATIONS

From the analysis of four-person fights some general observations begin to be made. It is frequently noted that there is always a pair of persons throwing pies at each other; this observation is readily accounted for by a variety of statements all of which amount to saying that there are always two people who are nearer to each other than any other pair.

The case in which two separate pairs throw at each other is commented upon as being two separate fights ('in different parts of the room') and sometimes generalized as always being a possibility when an even number of persons are involved in the fight.

The question of where to stand to avoid being hit is intuitively obvious from the start; a more interesting aspect eventually turns out to be the most dangerous place in the room.

FIVE-PERSON FIGHTS

The systematic method of checking for all possible four-person fights can be extended to a five-person fight. Splitting 5 into five numbers from the set $\{0, 1, 2, 3, 4, 5\}$ does not constitute much of a problem giving the following list of 7 ways.

$$(5, 0, 0, 0, 0) \ (4, 1, 0, 0, 0) \ (3, 2, 0, 0, 0) \ (2, 2, 1, 0, 0) \ (1, 1, 1, 1, 1)$$
$$(3, 1, 1, 0, 0) \ (2, 1, 1, 1, 0)$$

$(5, 0, 0, 0, 0)$ is then rejected as a model of a possible five-person fight for the same reason as $(4, 0, 0, 0)$ for a four-person fight. $(1, 1, 1, 1, 1)$ is not always readily rejected and often stimulates a good deal of argument before a good explanation of its impossibility is forthcoming.

Matching each of the remaining 'quintuples' to a corresponding fight-graph is straightforward and this can lead to a false conclusion if it is supposed that quintuples and fight-graphs are in one-to-one correspondence.

An alternative method of constructing by trial and error as many topoligically different fight-graphs as possible without reference to the quintuples reveals a flaw in this supposition in respect of $(2, 1, 1, 1, 0)$ and $(2, 2, 1, 0, 0)$ both of which describe two distinct graphs (the diagram shows those for $(2, 2, 1, 0, 0)$).

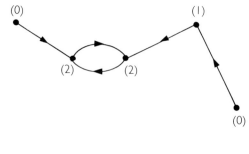

This adds a new dimension to the investigation and an alternative systematic method of determining all possible fight-graphs needs to be devised.

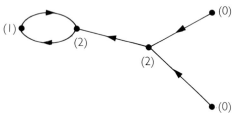

DRAWING FIGHT-GRAPHS SYSTEMATICALLY

Systematic methods of drawing can be devised with care. A frequently used strategy is to subdivide the graphs first according to the number of 'loops', where a loop occurs if two persons throw pies at each other. In a six-person fight there is just one case with three loops and so on. The other throws form chains attached to either end of some of the loops and can be added systematically. Nevertheless it proves all too easy to omit cases and group work helps here. At least 21 distinct six-person fights have been found.

LIVING DANGEROUSLY

At some stage the question of the maximum number of hits that can be received is often posed; this certainly arises if the systematic methods of the previous section are bypassed and fights between seven or more persons examined. A seven-person fight corresponding to $(6, 1, 0, 0, 0, 0, 0)$ proves impossible to draw. A regular hexagon is often drawn as shown in the diagram, with one person at each vertex and one at the centre and then bearing in mind the conditions of the problem that no two participants can be the

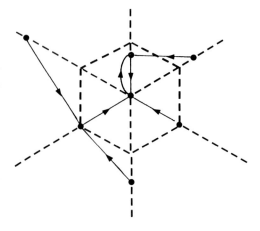

same distance apart an attempt is made to distort the figure slightly by moving five of the corners out from or in towards the centre by different amounts. This exercise always results in at least three of the outer ring being no longer nearest to the centre person. However a similar construction starting with a regular pentagon can leave all five corner persons firing at the centre.

Using the result about the longest side of a triangle being opposite the largest angle, it can be shown that the paths of any two pies hitting the same person must be at an angle greater than $60°$, and hence that a six-hit is impossible.

EXTENSIONS

The following are some questions which have been posed and which are worthy of exploration.

- It is a condition of the problem that where two people throw at each other the pies do not collide. Could any other two pies collide?
- If pies travel with equal speed, is the person hit most always hit first?
- Could one of the participants contributing to a five-hit also be the recipient of five hits (in, say, an eleven-person fight)?

Varying the rules

The rules of the problem can be varied; if each person is armed with two pies which are thrown at the two nearest persons then a three-person fight is bound to be $(2, 2, 2)$. Does such a triangular arrangement of throws inevitably occur as part of a fight with more than three participants?

Using incidence matrices

Enumeration of possible fight configurations for six or more persons has proved difficult. For older pupils and students a possible systematic method of determining all possible six-person fight-graphs could be by creating incidence matrices corresponding to the 'sextuples'. Such matrices *are* in one-to-one correspondence with the graphs; leading diagonal entries would be zero (self pie-infliction not allowed!), row totals would be 1 (each person throws one pie) and column totals would match the sextuple. Some graphs can be duplicated using this method but none should be missed.

REFERENCES

1 *American Mathematical Monthly*, April 1982, 274.

12   DESIGNING FIGURES

STATEMENT OF PROBLEM

This **L**-shaped figure has six sides and is therefore a hexagon. How many right angles does it have, five or six? Design a pentagon containing as many right angles as possible.

How many angles of 60° can a hexagon contain?

Design some figures similarly, to your own angle specifications.

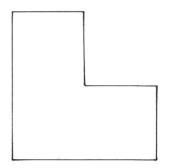

COMMENTARY

National Curriculum

A problem which forms part of this investigation is given as an example for level 6 of AT9. However the background knowledge required consists only of right-angles (level 2, AT10) and constructing 2-D figures from given information (level 4, AT10); at level 3 or 4 pupils can explore the problem, though systematic analysis of results is more likely level 6.

Materials

Squared paper is useful for early drawings but later its effect can be inhibiting of more adventurous experimentation with figures containing 'inclined' right angles. The same is true for triangled paper when considering 60° angles.

GENERAL COMMENTS

The problem of designing two- and three-dimensional figures to given specifications has always been a fascinating exercise. For example[1], pupils may be challenged to consider making a polyhedron with five triangular faces, or one with four quadrilateral and two triangular faces. The first is clearly impossible and the second possible but not easy to visualize. A later investigation, DESIGN (35) focuses on this type of 3-D problem.

A more recent article (Fielker, 1981)[2] in which the specifications are in terms of angles and therefore closely linked to the present investigation, discusses and extends the type of design outlined in the statement of the problem; the underlying philosophy in the article is strongly investigation oriented.

INITIAL APPROACHES

Attempts at drawing a pentagon containing only right angles, whether interior or exterior, invariably finish up with the first and fifth sides either along the same line, in which case a quadrilateral is formed, or parallel. Such a drawing is itself an explanation why more than three right angles is an impossible achievement.

The question of whether to explore both interior and exterior angles is an interesting decision. If only interior angles are considered to start with then the inclusion of exterior angles is an obvious extension. However some experimenters have found that to include both simultaneously is just as easy to handle because the two types of angle frequently complement each other.

Looking at a smaller number of sides is a useful strategy. The triangle is perhaps too trivial with its unique solution of one interior right angle but the quadrilateral is a little more interesting. Both figures are important if, eventually, a sequence of solutions is to be created. A quadrilateral can clearly have four interior right angles. On the other hand it cannot have just three interior ones, although two and one are possible. After a little consideration just one exterior right angle is also found to be a possibility.

PENTAGONS

Returning to the pentagon, the case with three interior right angles is often drawn on squared paper as in the diagram below, the two examples being different in so far as the first has the three right angles adjacent to each other and the second has not. The first can be seen as a square with a corner cut off, or a variation of the **L** shape in the statement of the problem which reduces the number of sides by one. This type of comparison of figures is helpful as one solution can lead to another and generalizations can be formed.

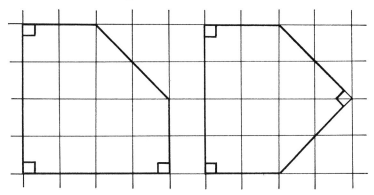

A solution for the case with two interior and one exterior right angle is derived directly from the second diagram by envisaging the isolated right angle pointing inwards instead of outwards, with minor adjustments to the lengths of the sides. Solutions can also be found for the cases with two interior right angles, with one of each type of right angle, and with just one right angle, of either type.

This method of describing solutions is considered unwieldy and either tabulation or an ordered-pair notation is soon adopted: (number of interior right angles, number of exterior right angles). Using the latter method, pentagon solutions can be found for the cases $(3, 0)$, $(2, 1)$, $(2, 0)$, $(1, 1)$, $(1, 0)$, $(0, 1)$ and trivially $(0, 0)$. Frequently all but the first two cases are regarded also as fairly trivial or insignificant, not being optimum possibilities. $(4, 0)$, $(1, 2)$ and $(0, 3)$ are impossible cases and so, for the pentagon, the maximum total number of right angles is three, the maximum interior number is three and the maximum exterior number is one.

AN EARLY GENERALIZATION

A frequent generalization is to extend the original **L** shape by creating additional steps; one additional step shown in the left-hand figure below gives a solution for eight sides with six interior and two exterior right angles, that is $(6, 2)$. Adding one more step gives a $(7, 3)$ solution for a ten-sided figure and so on. This sequence is one of the hints that suggest considering odd and even numbers of sides separately might be a good idea.

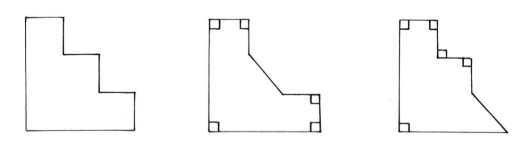

Cutting off one of the steps, as shown in the other two figures, gives either $(5, 0)$ or $(4, 1)$ as possible solutions for seven sides. Correspondingly, cutting off one step from a ten-sided staircase gives either $(6, 1)$ or $(5, 2)$ as possible solutions for nine sides.

TABULATING RESULTS

Results obtained so far clearly need tabulation; gaps in the following table show where results are as yet unresolved, and those entries with a question mark are hypothetical.

Number of sides	Maximum number of right angles			Optimal solutions*
	Total	Interior	Exterior	
3	1	1	0	(1, 0)
4	4	4	1	(4, 0) (0, 1)
5	3	3	1	(3, 0) (2, 1)
6	6	5?		(5, 1)?
7	5?			(5, 0)? (4, 1)?
8	8			(6, 2)
9	7?			(6, 1)? (5, 2)?
10	10			(7, 3)

*(x, y) notation represents (number of interior right angles, number of exterior right angles).

The hypothetical and unresolved cases require further experimentation with the hope of leading to appropriate generalizations. Nevertheless some patterns are evident already. The staircase generalization of the previous section clearly gives an optimum total number of right angles for the even-sided cases since *all* the angles in the figures are right angles; for the odd-sided cases could the corresponding sequence be 1, 3, 5, 7, . . . ? The 5 and 7 are, at present, hypothetical. However, because known solutions exist which give rise to them, it remains only for them to be exceeded. Furthermore some of these known solutions form a sequence (1, 0), (3, 0), (5, 0), but the solution (7, 0) for nine sides has yet to be found. If (7, 0) is possible, then is (9, 0) a possible solution for eleven sides, and so on?

CHECKING THE OPTIMAL CONJECTURES

Further investigation immediately seems necessary in respect of the six- and seven-sided cases. For six sides, there is little trouble in seeing that five is a maximum number of interior right angles, since any attempt to create a figure with *all* interior right angles inevitably falls at the last fence and forms an **L** shape. Solutions (5, 0) and (4, 1) are likewise found to be impossible but (4, 0) and (3, 1) are obtained by slight distortions of the fundamental **L** shape.

The question of whether two exterior right angles are possible with six sides is resolved when the solution (1, 2) is found. This often requires some new insight into the design process away from the **L** figure, one method being to modify the eight-sided staircase figure whilst retaining its two exterior right angles.

For seven sides, (5, 0) can be established as an optimum solution by endeavouring to create a figure with six successive right angles. Once five interior right angles have been drawn a sixth, either interior or exterior makes the seventh side parallel to the first. An optimum solution (0, 3) can be found for exterior right angles.

For nine sides, the (7, 0) solution can be found by creating a figure with seven successive right angles and joining up the last two points. Such an approach, however, fails to produce a (9, 0) solution for eleven sides *unless* sides of the figure are allowed to cross each other. This could introduce a new intriguing aspect into the investigation.

FURTHER DEVELOPMENT

Approaches from this stage onwards have been found to be distinctly individual at first, with a degree of random experimentation; it is a good point at which to dispense with squared paper. Eventually a systematic approach is adopted with a series of figures which can be generalized in the same way as the staircase sequence. Two examples of this are shown in the following diagram.

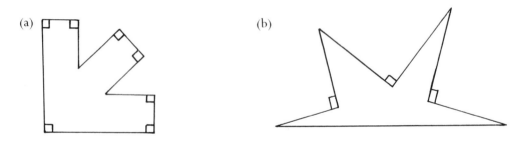

(a)   (b)

The 'square tooth' figure (a) is a (7, 0) solution for nine sides. With the addition of another 'tooth' it becomes (9, 0) for twelve sides and so on. The star shape (b) is a (0, 3) solution for seven sides to which further points can be added.

A THEORETICAL APPROACH

Older pupils and students normally approach the investigation with a mixture of experimental drawing and some theory derived from known interior angle sums of the polygons, preferably with the latter being expressed in terms of right angles. An exterior right angle counts as an interior angle of three right angles, for example, a pentagon has an interior angle sum of six right angles. Four interior right angles are impossible as the fifth angle would be two right angles or $180°$, making the pentagon a quadrilateral. There can be only at most one exterior right angle, as two such would absorb all the six right angles of the interior angle sum. This type of argument is then extended to other numbers of sides giving the following table of results. A-level students and beyond should be able to develop appropriate theoretical inequalities.

	Number of sides									
	3	4	5	6	7	8	9	10	11	12
Interior angle sum*	2	4	6	8	10	12	14	16	18	20
Interior maximum*	1	4	3	5	5	6	7	7	8	9
Exterior maximum*	0	1	1	2	3	3	4	5	5	6

*All figures are are in terms of right angles.

Shrewd observers readily point out that these theoretical results do not necessarily guarantee the existence of a corresponding figure. However no investigator has found evidence to the contrary so far.

One surprising aspect of this set of results is that, apart from some initial hiccups in the smaller numbers of sides, the pattern advances in steps of three and not as previously surmised in steps of two with even and odd numbers of sides forming different sequences.

GENERALIZING THE THEORY

The results have been generalized for n sides as follows $(n > 5)$.

Number of sides n, of form	Maximum number of right angles	
	Interior	Exterior
$3m$	$2m + 1$	$2m - 2$
$3m + 1$	$2m + 1$	$2m - 1$
$3m + 2$	$2m + 2$	$2m - 1$

For example, a 99-sided figure, in the form of a quarter of a cog wheel, could be drawn with 33 'square teeth' each tooth having three sides and two right angles, with one extra right angle at the bottom left-hand corner. The total number of interior right angles is $33 \times 2 + 1 = 67$, which accords with the theory. Further if the corner (right) angle is constructed on the other side of the teeth then all but three of the interior right angles become exterior and we have a 99-sided figure with 64 exterior right angles, which also accords with the theory.

This illustrates another surprising aspect, that maximum numbers of interior and exterior right angles *both* tend to the same limit of $\frac{2}{3}$ of the number of sides.

EXTENSIONS

Using $60°$ for the interior angles instead of $90°$ is mentioned in the statement of the problem; triangled paper is very helpful for experimental drawing initially. Some of the figures with large numbers of sides can become like Christmas trees, for example thirteen sides with a maximum of eight $60°$ angles and the other five angles each $300°$. Later, it will be necessary to dispense with the triangled paper as the number of sides increases.

A patterned sequence of drawings can be developed and likewise the theory based on interior angle sums, using $60°$ as the unit. The generalized results on this occasion advance in steps of five, instead of three as with the right angle.

Crossed and 3-D figures

Sticking with 90° angles, the possibility of overlapping figures has already been mentioned. Some care is needed in drawing to show where vertices do and do not occur; the first figure is an overlapping *hexagon* with the vertices marked by dots. Visualizing it in three dimensions so that there is no actual contact where the overlapping takes place might help. Does it have six interior right angles? How does one define interior in such figures?

Using the edges and face diagonals of a cube is another variation; the second figure shows a space pentagon with five right angles.

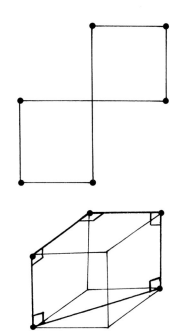

References

1 *School Mathematics Project Book 1* (1965) CUP, 115.
2 Fielker, D.S. (1981) Removing the Shackles of Euclid, *Mathematics Teaching*, September, 24ff.

13 SUMS AND PRODUCTS

STATEMENT OF PROBLEM

$1+2+3=1\times2\times3$.

Can you find another three numbers which have their sum equal to their product? If not, can you explain why?

Find some sets of two and three numbers such that their product is *twice* their sum.

Develop the investigation by varying some of the conditions, perhaps involving negative numbers or fractions.

COMMENTARY

National Curriculum

The only background knowledge required for substantial development of the investigation is addition and multiplication of integers, level 3 of AT3. However, a key feature is the need for a systematically organized search routine, which requires level 4 or 5. Random trial and error will not easily produce results, neither will it indicate whether all possible solutions have been found.

If negative numbers are to be included in the study then this requires at least level 5 (AT3). The use of algebra can feature strongly in the development; solving linear equations is level 6 (AT6) and algebraic generalizations of sequences of solutions implies level 7.

Materials

Squared paper is useful throughout, for recording and tabulating. Number rods give an opportunity to represent solutions geometrically.

GENERAL COMMENTS

This investigation started from the fact that there is only one pair of positive integers such that their sum and product are equal. For three positive integers, this is also the case but once variations are introduced, such as looking for sets of integers whose product is twice or three times their sum, the number of possible solutions and the relationship between them makes the problem quite intriguing.

INITIAL APPROACHES

Clearly pupils will focus on the positive integers at first, and just looking at pairs of these paves the way to an organized search.

Most pupils have little difficulty in picking up $2 + 2 = 2 \times 2$. However once the possibility of the product (P) being twice the sum (S) is included, a combined addition and multiplication table (abbreviated to S/P) has proved very helpful. Once compiled it proves to be a valuable resource throughout the investigation, even to the point where solutions involving fractions are considered.

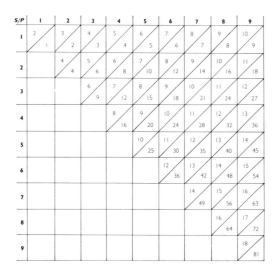

From the table it is easy to spot the pairs $(3, 6)$ and $(4, 4)$ for which $P = 2S$. We can also speculate whether the seond row (or second column) if extended would produce another such pair, and if not, why not.

Pairs $(6, 6)$, $(8, 8)$ satisfy $P = 3S$ and $P = 4S$ respectively, which suggests $(10, 10)$ will satisfy $P = 5S$ and so on to a generalization.

Extending the fourth row is a plausible possibility in seeking a pair for which $P = 3S$, and reveals the solution $(4, 12)$.

SETS OF THREE INTEGERS

The S/P table is also useful when moving on to three integers. Firstly if an explanation of the unique solution for $P = S$ has not been formulated then the table can provide one. If a search is made for three integers satisfying $P = S$ and one of the integers is 1, then the product part of the table is unchanged. So, seek pairs of integers in which the sum (S) is one less than the product (P). There is only one case, $S = 5$ and $P = 6$ giving $(1, 2, 3)$. Further, if none of the three integers is 1 then the lowest solution to be considered would be $(2, 2, 2)$ and this has $P > S$.

The table can also be used to find solutions for $P=2S$. Now we seek pairs of integers for which the sum (S) is one less than half of the product (P). It is not difficult to spot $S=9$, $P=20$ giving the triple $(1,4,5)$ and then $S=11$, $P=24$ giving the triple $(1,3,8)$ both of which are solutions for $P=2S$.

However, there may well be solutions for $P=2S$ in which none of the three integers is 1. The triple $(2,2,4)$ is usually discovered by first trying $(2,2,2)$ as an extension from $(2,2)$ for which $P=S$, for example by solving $2\times2\times x=2(2+2+x)$.

ILLUSTRATING SOLUTIONS GEOMETRICALLY

This proves quite useful; in some cases solutions can be found by this approach, using squared paper and/or Cuisenaire number rods, rather than searching the S/P table.

The diagrams below show some examples for sets of two and three integers.

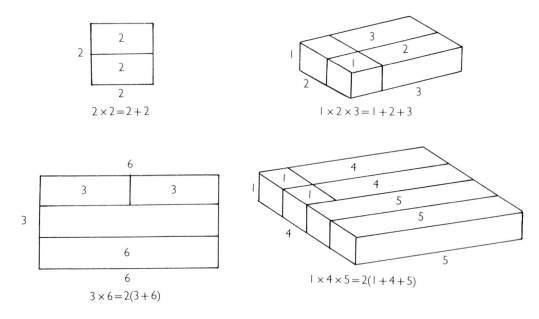

$2\times2=2+2$

$1\times2\times3=1+2+3$

$3\times6=2(3+6)$

$1\times4\times5=2(1+4+5)$

TOWARDS SEQUENCES OF SOLUTIONS

Having found three solutions for $P=2S$, a natural step is to consider $P=3S$. This time the S/P table can be searched for cases in which the sum (S) is one less than a third of the product (P). The triples $(1,6,7)$ and $(1,5,9)$ can soon be established. Collecting solutions so far suggests further ideas. The

Formula	Solutions		
$P=S$	$(1,2,3)$		
$P=2S$	$(1,3,8)$	$(1,4,5)$	$(2,2,4)$
$P=3S$	$(1,5,9)$	$(1,6,7)$	

(1, 2, 3) solution for $P=S$ has been so placed in the table to highlight a possible sequence of solutions. This sequence suggests (1, 8, 9) as a solution for $P=4S$ which is easily verified, and this is soon generalized by older pupils to $(1, 2n, 2n+1)$ for $P=nS$.

$P=S$	$2S$	$3S$	$4S$	$5S$	$6S$
1	1	1	1	1	1
2	3	4	5	6	7
3	8	15	24	35	48

Working on this principle of sequences of solutions (1, 4, 15) is found as another solution for $P=3S$, by continuing a possible sequence starting (1, 2, 3), (1, 3, 8) and solving $3(1+4+x)=4x$. For $P=4S$, the triple (1, 5, 24) is similarly found. The sequence $3, 8, 15, 24$ now has a clear 'rule' and tabulating leads to a clever generalization when it is observed that the integers in the third row are the product of the two integers either side in the row above. This identifies $(1, n+1, n(n+2))$ as a solution of $P=nS$.

MORE SOLUTION SEQUENCES

Another simple sequence is found by looking for solutions in which the smallest integer is 2, starting with (2, 2, 4) for $P=2S$, and continuing (2, 3, 5) for $P=3S$, (2, 4, 6) for $P=4S$ and generalized as $(2, n, n+2)$ for $P=nS$. It is interesting to represent these solutions geometrically, for example a cuboid of height 2, width 3 and length 5 can be built from two layers, the first of 3 rods of length 5, the second of 3 rods each of lengths 2 and 3.

Yet another sequence is found by writing the solution for $P=S$ as (1, 3, 2) and then continuing with the possible sequence (1, 4, 5), (1, 5, 9) from solutions already found for $P=2S$ and $P=3S$. Tabulating again helps and the proposed extensions to solutions for $P=4S$ and so on are easily verified.

$P=S$	$2S$	$3S$	$4S$	$5S$	$6S$
1	1	1	1	1	1
3	4	5	6	7	8
2	5	9	14	20	27

This sequence proves much more difficult to generalize simply on the basis of observation. Analytically, it can be shown to be $(1, n+2, \frac{1}{2}n(n+3))$ for $P=nS$. However quite a lot of interesting speculations arise from observing that the first three cubes $1, 8, 27$ have the property $P=6S$. This is linked to the basic solution (1, 2, 3) for $P=S$ through the result

$$1^3 + 2^3 + 3^3 = (1+2+3)^2$$

which is well known to students of A-level mathematics.

A SUMMARY OF SOLUTIONS SO FAR

Those so far discovered can be tabulated.

Formula	Solutions			
$P = S$	$(1, 2, 3)$			
$P = 2S$	$(1, 3, 8)$	$(1, 4, 5)$	$(2, 2, 4)$	
$P = 3S$	$(1, 4, 15)$	$(1, 5, 9)$	$(1, 6, 7)$	$(2, 3, 5)$
$P = 4S$	$(1, 5, 24)$	$(1, 6, 14)$	$(1, 8, 9)$	$(2, 4, 6)$
$P = 5S$	$(1, 6, 35)$	$(1, 7, 20)$	$(1, 10, 11)$	$(2, 5, 7)$
$P = 6S$	$(1, 7, 48)$	$(1, 8, 27)$	$(1, 12, 13)$	$(2, 6, 8)$

This list is, as one might possibly expect, far from complete with $(2, 2, 12)$ and $(3, 3, 3)$ missing from $P = 3S$ and so on. How could we ensure that it is complete?

COMPUTING

An alternative approach to sequencing solutions is to write a simple program in BASIC with three nested FOR NEXT loops which searches exhaustively for solutions to each case through a limited range of integers. Solutions so found can then be linked by the sequence methods discussed in the previous sections.

INVOLVING NEGATIVE NUMBERS

If negative integers are allowed then a new solution can soon be found for $P = S$ by supposing that one integer in the solution is 0. P is then 0 and the other two integers could be 1 and $^-1$. Of course $(2, 0, ^-2)$ would do as well, and generally $(n, 0, ^-n)$.

The solution $(^-1, ^-2, ^-3)$ to $P = S$ is found by extending the computer search of the previous section and this rapidly leads to the general result that if (a, b, c) is a solution to any of $P = nS$ in positive integers then $(^-a, ^-b, ^-c)$ is also a solution.

A neat sequence of solutions which can be found from the computer search is tabulated here and has the easy generalization $(1, n-1, ^-n^2)$ for $P = nS$.

$P = S$	$2S$	$3S$	$4S$	$5S$	$6S$
1	1	1	1	1	1
0	1	2	3	4	5
$^-1$	$^-4$	$^-9$	$^-16$	$^-25$	$^-36$

EXTENSIONS

Solutions involving fractions

These prove quite difficult at first, until a connection between the solution $(2, 4, 6)$ for $P = 4S$ and $(1, 2, 3)$ for $P = S$ is found. Various sensible reasons can be produced to explain that if (x, y, z) satisfies $P = S$ then $(2x, 2y, 2z)$ satisfies $P = 4S$.

This result immediately establishes the triples $(\frac{1}{2}, 2\frac{1}{2}, 12)$, $(\frac{1}{2}, 3, 7)$, $(\frac{1}{2}, 4, 4\frac{1}{2})$ as solutions of $P = S$ from the already known solutions of $P = 4S$.

These perhaps surprising results call for some reflection and checking back. Could they have not been found by a simpler method? If a triple for $P = S$ has the form $(\frac{1}{2}, x, y)$ then $1 + 2(x + y) = xy$; the S/P table can be searched again, rapidly revealing $x = 3$, $y = 7$ as a possibility. This approach could be developed further.

Another development which leads to fractional solutions is a sequence of solutions all of which have at least one number 3. By interpolation, $P = 2S$ has a solution $(3, 2\frac{1}{2}, 2)$, likewise for $P = 4S$ and so on.

Allowing negative numbers and fractions reveals the simple sequence of pairs $(\frac{1}{2}, -1)$, $(\frac{1}{3}, -\frac{1}{2})$, $(\frac{1}{4}, -\frac{1}{3})$, . . . all of which satisfy $P = S$.

$P = S$	$3S$	$5S$	$7S$	$9S$
3	3	3	3	3
2	3	4	5	6
1	3	5	7	9

Sets of four integers

Extending the problem to four integers is more straightforward. For $P = S$ if three of the integers are 1 and the fourth x then $S = 3 + x$ and $P = x$ which is impossible. If however two of the integers are 1 and the other two x and y then $S = 2 + x + y$ and $P = xy$ and the S/P table can be used to find sums (S) 2 less than the product (P), which occurs for the pairs $(2, 4)$ and $(4, 2)$. So $(1, 1, 2, 4)$ is a solution for $P = S$.

This can be extended to a sequence $(1, 1, 4, 6)$, $(1, 1, 6, 8)$ for $P = 2S$, $P = 3S$, with the generalization $(1, 1, 2n, 2n + 2)$ for $P = nS$.

The computer search method can be used and another sequence soon found is $(1, 2, n, n + 3)$ for $P = nS$.

14 CRAZY PAVING

STATEMENT OF PROBLEM

A paving stone is a rectangle 2 units long and 1 unit wide. The diagram shows a paving of a square of side 4 units using eight stones. It is 'faulty' or not 'completely crazy' because it has two straight 'fault' lines, AB and CD, which run from side to side through it.

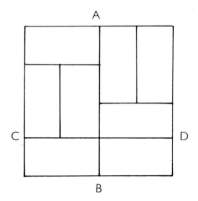

Can you rearrange the paving stones in the square so that there are less than two faults?

Investigate paving various sizes of rectangle, each time trying to make the number of fault lines as small as possible.

COMMENTARY

National Curriculum

Creating patterns with 2-D shapes is level 2 (AT10). However, the strategies and reasoning involved in exploring the fault lines requires at least level 4.

Materials

With younger pupils, Cuisenaire number rods can be usefully employed and even with older pupils and students a visual aid, such as dominoes, to represent the stones can be very helpful.

Squared paper is clearly a valuable help for recording.

GENERAL COMMENTS

This investigation is built around the problem, which may well be familiar, of finding completely faultless rectangles paved with 2-by-1 rectangles. Experience of discussing this problem with pupils and students has shown firstly that a great deal of trial and error is involved with often only a slight chance of success and secondly that, frequently, the types of argument which explain the possibility or impossibility of paving particular sizes of rectangle faultlessly are complex.

However, introducing the counting of fault lines with the aim of minimizing these makes the problem more generally tractable. In the case of small rectangles all possible pavings can be determined and classified according to the number of fault lines.

The early stages of the investigation have a good deal in common with the investigation WALLS (5). However, in the present context all the possible arrangements of stones to form a required paving are not necessarily classified as different. The two 2-by-5 pavings in the diagram below are in principle the same, each being made from three blocks of type A and one block of type B laid side by side and hence always giving rise to three fault lines across the paving.

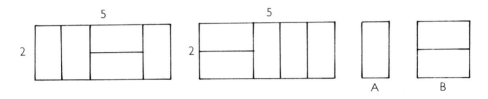

The most regular arrangement of stones in a pathway usually gives rise to a maximum number of fault lines. The issue however is not always clear-cut and this aspect is also worth including in the investigation.

INITIAL APPROACHES

Paving 'pathways' which are 1 unit wide and of even length is straightforward with each case being unique and the number of faults one less than the number of stones.

Pathways which are 2 units wide are not so straightforward but after some experimentation the results in the following table can be established.

Number of fault lines	Path length									
	1	2	3	4	5	6	7	8	9	10 . . . n
Minimum	0	1	1	2	2	3	3	4	4	5 . . . $[\frac{1}{2}n]$
Maximum	0	1	2	3	4	5	6	7	8	9 . . . $n-1$

The maximum case is again straightforward with the stones being laid lengthwise side by side with one less faults than the number of stones. The results for the minimum case can be explained although for even length paths essentially different arrangements give the same minimum as in this diagram. All paths, however, are effectively made from the same two types of 'blocks' referred to in the general comments section, the minimum being achieved by using as many B blocks as possible.

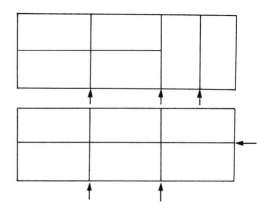

WIDER PATHS

With paths 3 units wide, a new aspect is introduced. It is often suggested at first that all 3-unit width pathways can be constructed from the basic blocks shown in this diagram. However, if a 3-unit width pathway is made by laying some of the 1- and 2-unit width pathways side by side then a counter example

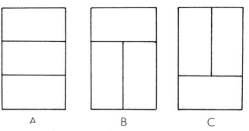

will be found. Consequently the minimum number of faults seems to be one in all cases and the maximum is once again straightforward using blocks of type A only.

With 4-unit width pathways, systematic classification of the range of cases involved is not so easy. The maximum case shows the first sign of not being simple; for the minimum case a helpful strategy is to try to block all potential fault lines as the path is laid (see the following section), though this is found to be an impossibility; the minimum appears to oscillate between 1 and 2 faults, for odd and even path lengths respectively.

Tabulation of the results obtained so far is shown in the following table, a dash (–) being entered when the paving is not possible (odd by odd).

Minimum number of fault lines

Width	\ Length 1	2	3	4	5	6	7	8	9	10
1	–	0	–	1	–	2	–	3	–	4
2	0	1	1	2	2	3	3	4	4	5
3	–	1	–	1	–	1	–	1	–	1
4	1	2	1	2	1	2	1	2	1	2

Maximum number of fault lines

Width	\ Length 1	2	3	4	5	6	7	8	9	10
1	–	0	–	1	–	2	–	3	–	4
2	0	1	2	3	4	5	6	7	8	9
3	–	2	–	3	–	4	–	5	–	6
4	1	3	3	4	5	6	7	8	9	10

By symmetry, the numbers in the first four columns can be extended in line with the first four rows. The pattern evident in the 'maximum' table gives scope for predicting how the other rows and columns extend, but in the 'minimum' table, beyond an apparent decreasing trend (with increasing width) in the number of fault lines when the length is greater than 5, confident predictions appear to be difficult. The following section outlines an approach which overcomes this problem.

BLOCKING FAULT LINES

Pathways can be laid progressively with the strategy of blocking all potential fault lines as they occur. With pathways of width two it immediately becomes obvious that faults across the pathway cannot be avoided. With width three, a sequence of stages in laying the pathway is shown in the following diagram starting with one stone lengthwise across the path and one lengthwise along the path.

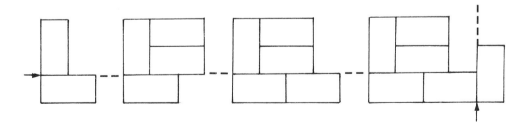

This shows that in attempting to avoid faults across the path a fault along the path is inevitably extended. Likewise, if this fault is blocked a fault across the path is automatically created.

Pupils produce arguments which are equivalent to this but theirs are often more lengthily stated.

Similar sequences can be developed for pathways of width four; all have to start with one stone laid lengthwise across the path and two lengthwise along the path.

With width five, a start can be made with two stones across the path and one along, or one stone across and three along the path. With the first of these it is possible to produce a completely crazy, that is fault-free, path of length six, though not with the second for which a length of eight is necessary for a fault-free result. Explanations of this do not appear easy to come by, unless the avenues suggested in the next section are explored.

ANALYSING BLOCKAGES

Pavings can be examined for the way in which possible fault lines are blocked by stones. Several different pavings of the same size rectangle can be analysed in the manner of this diagram, which shows two pavings of a 4-by-5 rectangle with the number of stones blocking each possible fault line indicated by the arrows. If this is done for different sizes of rectangle, some consistency in the numbers of blocking stones can be discerned. Pupils raise the following type of questions and frequently resolve them successfully.

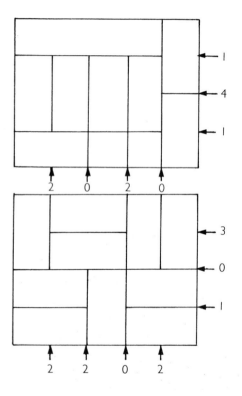

- Why is the total of blockings (for example, 10 in the diagram) equal to the number of stones in the paving?
- Why are some possible fault lines blocked by an even number of stones and others by an odd number of stones?
- Along which lines can faults never occur?

It is possible now to argue for the impossibility of a fault-free 4-by-5 rectangle. The four vertical lines in the diagram above would each have to be blocked by at least two horizontal stones, and the three horizontal lines each by at least one vertical stone; but, there are only *ten* stones in the paving . . .

EXTENSIONS

Shapes other than rectangles can be considered for paving. One neatly organized fault-free paving is a pathway of width 2 units around a square 'pond' of side 1 unit. This idea can be generalized, for example, for the same width paths round any rectangular pond of odd dimensions.

Pavings of rectangles, in which at least one dimension is a multiple of three, with 3-by-1 stones can be analysed along precisely the same lines as the approaches described for 2-by-1 stones. If sufficient data is collected for tables like those found for 2-by-1 stones, patterns in the numbers will be more strongly apparent. It is hypothesized that the smallest fault-free rectangle in this case is 9 by 7.

15 CLONES

STATEMENT OF PROBLEM

Four squares fit together to make another square of twice the dimensions and four equilateral triangles do likewise.

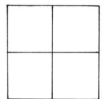

Can you find other figures which have this cloning property, such that when a number of them are fitted together a larger version of the same figure is formed?

Consider instead a number of squares of *differing* sizes fitting together to form a square, and extend this idea to other shapes.

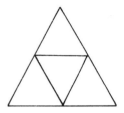

COMMENTARY

National Curriculum

Ratio, tessellation, enlargement and **similarity** are strong features of this investigation. When fully developed it should therefore draw on and give good experience of level 6 in AT2 and AT11. However pupils at earlier levels can explore some of the simpler cases using gummed shapes or templates.

Materials

Squared and triangled paper are essential for much of the study.

GENERAL COMMENTS

There are a number of special cases of figures which function in the required manner and there is a danger that the investigation can become a collection of relatively unrelated examples. The approaches described in the following sections therefore do not form an overall systematic line of development as with many other investigations. Young pupils can explore some simple cases, but older pupils and students should attempt to sustain an organized enquiry into one or two of the approaches suggested.

The investigation can be seen as a subset of the wider problem of tessellating figures. If a figure clones itself then clearly the process can be repeated and the figure therefore can be the unit of a tessellation. The converse is not true, since a figure that tessellates does not necessarily clone. A good example is the regular hexagon; it is impossible to outline, using the edges of a tessellation of regular hexagons, a larger hexagon similar to the unit hexagon forming the tessellation.

The inverse approach to that suggested in the statement of the problem, i.e. that of dissecting a shape into similar smaller ones, is a useful method of attack in some cases.

INITIAL APPROACHES

A 2-by-1 rectangle is a good starting point. There are five ways of fitting *four* of these together to form a 4-by-2 rectangle (but only three ways if symmetrical pairs are counted as one way). With *nine*, the number of possible ways of making a 6-by-3 rectangle is clearly much larger. It seems impossible to use any other than a square number of these rectangles to form a larger one with a length to width ratio of 2 : 1. It is certainly worth challenging pupils to explain why it is impossible with three, and perhaps with five.

The case where four 2-by-1 rectangles placed side by side make a 4-by-2 rectangle prompts the question of what dimensions has a rectangle to be in order that three fit together to clone themselves as in the left-hand diagram below.

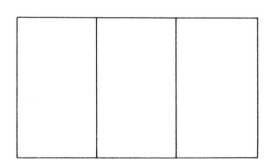
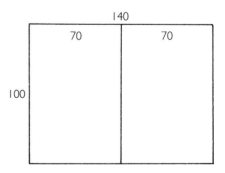

The right-hand diagram shows an attempt to find a rectangle which clones itself when just two are used. Suitable dimensions in millimetres could be found by experiment using squared paper and a calculator. In the diagram, two 100-by-70 rectangles are fitted together to form a rectangle, 140-by-100. Now $140/100 = 1.4$, and $100/70 = 1.43$. While these are not quite the same, just a small adjustment is now necessary to get more precise proportions of the required rectangle.

We are in the realm of standard paper sizes. A good test of the correctness of the proportions of an A4 sheet is to fold it in half (to make an A5 sheet) and place it corner to corner on another sheet of A4 as shown in this diagram. If the proportions are correct then the diagonals from the common corner of the A5 and A4 should be superimposed. This test should be applied to a rectangle of the proposed dimensions.

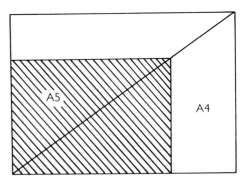

SQUARE-BUILT CLONERS

Figures composed of squares provide another line of approach. The case of three squares in an **L** shape as shown here is well known but there are many others which can be searched for, in the first instance by experiment on squared paper.

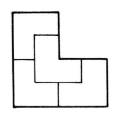

It is important to explain why it appears impossible to clone certain figures as well as to find successful examples. Of the four **L** shapes in the next diagram, three will clone using four at a time and one will not.

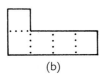

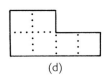

(a)   (b)   (c)   (d)

Successful cases can be used to find others, for example, figure (d) above clones because it is a transformation of the standard **L** shape stretched by a factor of two in one direction. Such a transformation however fails with the other two successful 'cloners' in the above set of four. Why should this be?

Further, if a clone is possible with four figures, then is it also possible with nine at a time? If so, is the converse true?

A GENERAL THEORY

A step towards generalization was made by a student who proposed the following hypothesis:

'If two identical figures, made up of squares, can themselves be fitted together to form a square then cloning is possible. The number of figures required to form the large version is twice the number of squares in the basic figure.'

This is illustrated in the diagram. In the smaller figure, two **F** shapes are each formed from eight squares, and fit together to form a square. Eight such squares then form a large **F** which is equivalent to the small one scaled up by a factor of four.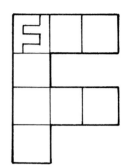

Clearly there are many more examples of such shapes formed by dividing a 4-by-4 square through its centre point into two identical parts each consisting of eight squares.

A DEVELOPMENT FROM SQUARES

One development is to involve right-angled isosceles triangles (with angles 90°, 45° and 45°). This shape clones successfully using four figures since it can be viewed as half of the standard **L** shape made up of three squares cut by its one line of symmetry. Drawing the line of symmetry through the cloned larger **L** (see the diagram on page 108) then shows how the cloning takes place.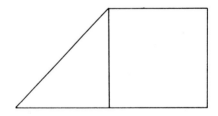

Further development looks a plausible possibility but other examples of combinations of squares and one or more of these triangles has yet to be found to work.

TRIANGLE–BUILT CLONERS

Combinations of equilateral triangles can be investigated along similar lines to that of combination of squares. The rhombus formed by two triangles clones in the same way as a square. The trapezium (half of a regular hexagon) formed by three triangles also successfully clones itself as shown here. This shape is a frequent one for tables in classrooms, giving a useful visual aid.

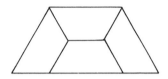

Following the same principles of the previous section, the line symmetry of the trapezium gives another cloner consisting of one and a half equilateral triangles.

The general theory about square-based figures can be applied to equilateral triangles. This diagram shows an example in which each of the three figures interlocking to form the large equilateral triangle is formed from 12 small triangles and hence 36 figures clone one 6 times the size.

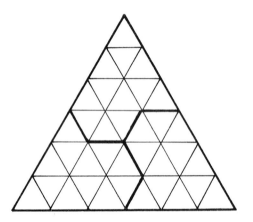

One problem in handling triangular-based figures is that identical ones in different orientations are not always as simple to pick out as in the case of squared-based figures.

FURTHER IDEAS

All triangles clone following the same principle as the equilateral triangle and likewise the cloning of squares generalizes (by an affine transformation which preserves ratios along lines) into parallelograms. The right angled isosceles triangle (with angles $90°$, $45°$ and $45°$) is a particular case in which cloning can be achieved with 2, 8, 18, 32, . . . triangles as well as with the usual 4, 9, 16, ... Further development of cloning triangles is outlined in the next section.

EXTENSIONS

Relaxing the constraint of keeping all the figures congruent is mentioned in the statement of the problem.

Squares into squares

Subdividing squares into integral-sided smaller squares is worth exploring. The investigation can start with a 3-by-3 square which can be split into either nine unit squares, or one 2-by-2 square and five unit squares. A 4-by-4 square can be split into either 4, 7, 8, 10, 13, or 16 smaller integral-sided squares in various size combinations. What about a 5-by-5 square?

New cloning figures

Some figures which previously did not clone now do, for example the $60°$, $90°$, $120°$, $90°$ large kite in this diagram; all the smaller kite shapes are similar to the large one.

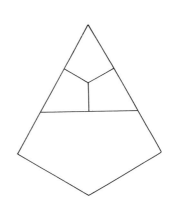

Any right-angled triangle can be split into two non-congruent triangles both similar to the original, by drawing the altitude from the right angle vertex, and this process can be continued indefinitely.

An unusual result

A student proposed an example in which four similar but not all congruent triangles clone a similar triangle, as shown in the diagram. ABC is the first triangle and a congruent one is fitted to it to produce the parallelogram ABDC. If the dotted line XY is now drawn parallel to BC then the standard figure of four congruent triangles cloning AXY is obtained. If however PQ is drawn through D so that $\angle AQP = \angle ABC$ then triangle APQ is similar to all its four constituent triangles.

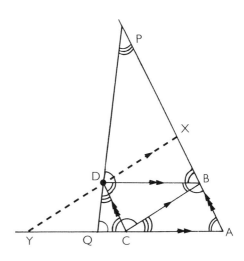

16   THIRDING A SQUARE

STATEMENT OF PROBLEM

On squared paper draw a square of side 4 cm, mark the centre point of the square and draw a straight line through this point. Does this line divide the square into two equal parts? Does *any* straight line through the centre halve the square?

Now draw a square of side 6 cm and mark a point (X) half-way across and one-third the way down the square as shown. Then draw a straight line through this point. Does this 'third' the square, that is cut off one-third of it? Does *any* straight line through this point third the square?

Investigate lines, straight or otherwise, which halve or third squares and other figures.

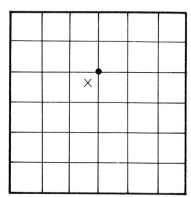

COMMENTARY

National Curriculum

This would be a good investigation to link with developing the concept of **area**. Area by square counting is level 4 of AT8, but use of formulae is as high as level 8. All that is necessary here is to be aware that diagonals halve the area of rectangles and parallelograms.

Congruence (level 5, AT10) is also relevant, but the ideas of congruence involved here are probably earlier than that level. **Symmetry**, both reflective and rotational (levels 3 and 4, AT11) is also involved.

Materials

Squared and triangular paper are invaluable throughout the investigation.

GENERAL COMMENTS

The title of the investigation may recall one of the three renowned 'impossible' ruler and compass construction problems, that of trisecting an angle (the other two being duplicating a cube and constructing a regular heptagon).

The problem of dividing figures into two or more parts of equal area or volume, but which may or may not be congruent parts, has also received frequent attention in mathematics and is much more amenable to exploration. A specific case is the fascinating 'ham sandwich' theorem which states that with a single plane cut, it is always possible to slice both pieces of bread and the ham inside, no matter of what irregular shape they may be, each into two parts of equal volume. The proof of the theorem establishes only the existence of such a possibility, without specifying where the cut should be made in a particular instance!

In this investigation, pupils and students with a knowledge of GCSE level mathematics, and beyond, frequently resort to using formulae for the area of a triangle, instead of employing its relationship to appropriate rectangles or parallelograms, and as a result their resolutions of the various problems can become unnecessarily complex and unwieldy. Too much knowledge sometimes can be a serious disadvantage.

INITIAL APPROACHES

Young pupils can investigate how to divide a 4-by-4 square on squared paper into two congruent parts using only grid lines; one such method is shown here. (If this approach is adopted then the statement of the problem should be reworded.)

Deciding which methods are symmetrically in principle the same, and whether all possible methods have been found gives rise to a lot of speculation and discussion.

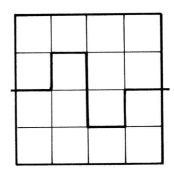

This can be extended to a 6-by-6 square which clearly has many more solutions.

Further, a slight relaxation of the grid-line restriction by allowing one diagonal of a square to be used, brings odd-sided squares within consideration.

WHERE TO DRAW THE LINE

All the dividing lines in the previous section pass through the centre point of the square as the statement of the problem suggests. Pupils have little difficulty in formulating their own sensible explanations of why any line through the centre halves the square but find it comparatively difficult to explain cogently why *only* such lines divide the square into two congruent parts. Looking at the question of dividing the square into two non-congruent parts of equal area usually generates little interest, largely because it is a too unrestricted problem.

In the diagram in the statement of the problem, thirding the square clearly occurs with a line through X parallel to one pair of sides and, equally clearly, a line through X parallel to the other pair of sides halves the square. The line drawn in this diagram is

soon realized to third the square also, by referring to the equality of the triangular parts above and below the line parallel to two of the sides. It is then suggested that the line through X can be rotated further until it reaches a corner of the square, in which position the line halves a rectangle which is two-thirds of the square. The solution to a halving problem has helped in the resolution of a thirding problem, a feature which illustrates how an easier analogous problem can help to resolve a more difficult one.

Further rotation of the line causes difficulties.

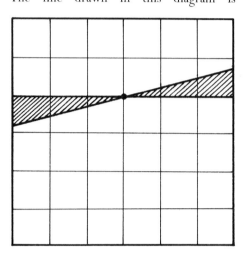

A FAMILY OF 'THIRDING' LINES

Frequently at this point it is realized that there are three other points which behave like X and a diagram like the one below is drawn with all the internal lines being 'thirding' lines. Discussion then focuses on the critical cases of the lines through the corners. Rotating these

in one direction so that they cut opposite sides of the square preserves their thirding property but doubt is expressed about rotating them in the other direction so that they cut adjacent sides of the square. Does joining and extending XY, for example, third the square? There is little difficulty in finding that it cuts off 12.5 squares (halving a 5-by-5 square) instead of 12.

Attention then focuses on the point P because a line through this point cutting adjacent sides would clearly cut off less than the line through X and Y. Resolution of the problem often sticks at this point.

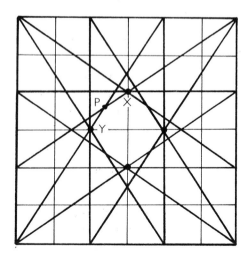

CUTTING OFF ONE-SIXTH

Switching instead to a consideration of lines that cut off one-sixth of the 6-by-6 square can bring more progress when coupled with further use of the insight which views the lines as halving appropriate rectangles. The diagram here shows twenty such 'sixthing' lines.

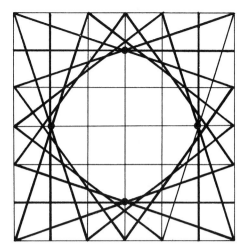

Although three of these lines pass through each of the four marked points there is clearly a curve-stitching or envelope effect being produced as well. Those with some knowledge of conic sections will recognize the property of a hyperbola whereby a tangent forms a triangle of constant area with the asymptotes.

RECTANGLES AND PARALLELOGRAMS

Though it is obvious to all pupils that the two diagonals of a square quarter the square, it is worth asking them to explain why the same is true for the diagonals of a rectangle and a parallelogram as the four triangles are no longer all the same shape. This result is useful if the problem of halving a triangle is tackled.

Thirding a rectangle follows similar principles to that of thirding a square and this slight generalization causes little difficulty. The case of a parallelogram seems more of a problem at first until one considers slicing it into three equal parts by pairs of lines parallel to either pair of parallel lines. These together divide the parallelogram into nine equal parallelograms similar to the original, and thirding lines through the corners of the parallelogram can now be drawn.

THIRDING AN EQUILATERAL TRIANGLE

The case of the triangle can be more interesting and challenging. Starting with halving an equilateral triangle is straightforward at first with three lines each joining a corner to the middle of the opposite side, the medians of the triangle. This can be generalized for any triangle and in this context it is helpful to view the triangle as half of a parallelogram in three different ways. The quartering property of the diagonals of these parallelograms mentioned above demonstrates the halving of the triangle by its medians.

Returning to the equilateral triangle, other lines through the centre point do not halve the triangle as pupils often confirm by drawing a 6-cm-side triangle on triangled paper. A line through the centre point parallel to one of the sides is found to cut off only 16 of the 36 small triangles.

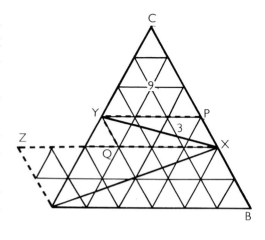

Trial and error experimentation to find what part of the triangle is cut off by joining other specific points of the sides is worth encouraging. Triangle counts are quite feasible when appropriate half-parallelograms are considered. In the diagram here AX, when seen to be halving the parallelogram ABXZ, is found to cut off 12 small triangles and hence third the triangle. XY can also be found as a thirding line, either through halving parallelogram XQYP and thus cutting off 3 plus 9 small triangles as shown, or by halving triangle AXC by a median.

Some pupils could be challenged to draw a set of lines which cut off one-sixth of a 12-cm-side equilateral triangle. A similar envelope effect to the diagram in the section on cutting off one-sixth will be produced.

EXTENSIONS

Quartering a square

The two diagonals of a square cut at right angles and quarter the square. If the two diagonals are rotated about the centre point so that they remain at right angles (but are no longer diagonals) do they still quarter the square?

Does a similar result apply for the diagonals of a rectangle? If not, how are two quartering lines through the centre point to be drawn? Consider the two extreme cases of the diagonals and the two lines parallel to the sides. Where would 'half way between' these two cases be?

Cutting other plane figures

Halving and thirding other plane figures than those already considered needs some care. Halving an even-sided regular figure is straightforward and thirding a regular hexagon with two lines through a vertex has a justifiable intuitive solution.

Older pupils and students with some geometrical background could look at halving a quadrilateral by a line through a vertex.

Halving two figures at one go

Two squares in a plane are simultaneously halved with a single straight cut through their centres. Would the same be true for two rectangles?

Could a single cut, simultaneously, halve one square and third the other?

Can a general principle for constructing a line which simultaneously halves two triangles be developed?

17 FIGURE SEQUENCES

STATEMENT OF PROBLEM

Here is a square within a square. The smaller one is formed by joining up the middle points of the sides of the larger one. What happens if you join up the middle points of the sides of the smaller one in the same way? and so on ...?

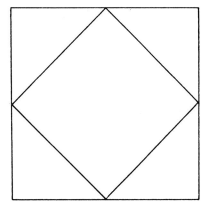

Start instead with a triangle, and then a quadrilateral, again using middle points. Comment on the shapes of the inside figures.

Experiment with other starting figures and with different positions on the side, for example, one-quarter of the way along each side.

COMMENTARY

National Curriculum

The concept of a **limit** is a fundamental one in mathematics and is a significant feature of this investigation. Here we have the visual appeal of diagrams to strengthen experience of the concept. There is no explicit use of the word limit in the National Curriculum though it is implicit in the higher levels of ATs 5 and 7.

Other geometrical concepts involved are those of **parallels, area, ratio** and **similarity** all of which suggest a minimum of level 6.

A reasonable degree of skill in accurate drawing and in mental calculation is also required if the investigation is to be developed to any depth.

For older pupils, **vectors** (level 8, AT11) are particularly useful for resolving observations about the types of polygon to which the sequences appear to converge.

Use of computer **graphics**, mentioned in ATs 7, 10 and 11, creates some striking effects with much less labour than hand drawings (see the General Comments section).

Materials

Plain, squared and triangled paper are all used at appropriate stages.

GENERAL COMMENTS

Originally this investigation arose from considering duality in polyhedra. If, for example, the centre points of adjacent faces of a cube are joined up then an octahedron is formed within the cube. If the process is repeated with the octahedron then another cube is formed and so on.

The two-dimensional analogue is easier to handle and a wide range of cases is readily available for exploration.

For a computer graphics program there are two fundamental procedures:

1. to draw a figure given the coordinates of its vertices;
2. to calculate the mid-points (or some other proportion) of the sides, which are then substituted for the original ones and the procedures repeated.

Either LOGO or BASIC can be used.

INITIAL APPROACHES

Squared paper should be used for drawing and extending the figure in the statement of the problem. A square of side 8 cm is a good starting one, since the vertices of the first six squares in the sequence are easily located at vertices of the square grid. Pupils could be asked to find the area and length of side of each square in the sequence as a proportion of the initial square.

To explore the quarter-point case, start with a square of side 16 cm which fits neatly onto A4. The second figure of the sequence is easily drawn and it will come as a surprise to many pupils that it is also a square. They should be challenged to explain why it is a square and to find its area. Conveniently the vertices for the next square will be found to be at vertices of the square grid. This third square will be observed to have a side of 10 cm which can be explained from the area ratio of each square to its predecessor being 5/8. The vertices of the fourth square also lie on grid lines.

With care, further squares can be added to the diagram. Accuracy is more likely to be ensured if at each stage all three points along each side which divide it into four equal parts are marked. The technique most frequently employed is to measure the length of side in millimetres, mentally halve it to mark the centre point and then mark the two quarter-points similarly.

Fortunately the convergent nature of the limiting process ensures that small errors are to some extent self-correcting.

PURSUIT CURVES

If sufficient squares have been drawn in the quarter-point case then the spiralling pattern of squares will have produced an effect of four curves, one from each corner, spiralling towards a limit point. This effect is much clearer if a smaller fraction is used such as 0.1, and although the drawings take some time to produce, the end result is very pleasing as shown in this diagram.

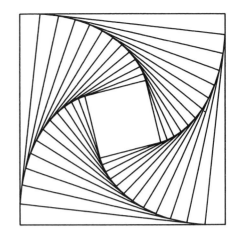

The effect is even more striking if several pupils produce drawings which can be cut out and formed into a tessellation for display.

The spirals are known as **curves of pursuit**. If four pupils stand at the corner of a square ABCD and at given signals each pupil moves one stride towards the next pupil A to B, B to C, C to D, and D to A then each pupil will follow a spiral curve of pursuit and eventually all finish up together at the centre of the square. This physical modelling of the process suggests a slightly easier method of construction of such curves. At each stage mark a point on each side the same distance (for example, 1 cm if the initial square side is 10 cm) from the corner of the square to form the vertices of the next square. The ratio between each square and the next will no longer be constant but the overall effect is the same.

OTHER FIGURES

The themes of the previous two sections can be applied to other familiar geometrical figures. Starting with a rectangle the middle-point case gives a sequence of alternate rhombuses and rectangles each one half the area of its predecessor. An equilateral triangle start results in a sequence of such triangles each one-quarter the area of its predecessor and alternately point up and point down.

A scalene triangle follows the same pattern as the equilateral triangle. When the first mid-point triangle is drawn the figure consists of four congruent triangles similar to the starting triangle. It is at this stage that the theorem traditionally known as the **mid-point theorem** begins to play an important rôle in further development, particularly if observed results are to be analysed and explained. ('The line joining the mid-points of two sides of a triangle is parallel to the third side and equal to half of it.')

THE QUADRILATERAL CASE

With a quadrilateral start, a striking result
will be observed. The second quadrilateral of
the sequence is always a parallelogram as
shown in this diagram, and the succeeding
figures are also parallelograms of two different shapes. The area ratio is constantly one-
half as with the square and rectangle cases,
and this is simple enough to explain for the
sequence of parallelograms. However, the
first stage from quadrilateral to parallelogram is not obvious.

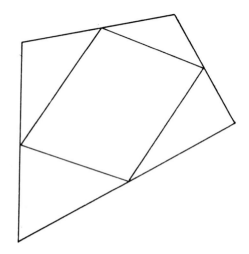

    Pupils have demonstrated the result by
dissection. First the quadrilateral is cut out and then the four triangular corners are cut off
and rearranged to cover the remaining parallelogram. This can be done in either of two ways,
but once achieved a diagram should be drawn showing how, and possibly why, the dissection
works.

    A variety of theoretical explanations are possible, all similar in principle. One of the
neatest involves drawing pairs of parallel lines to the diagonals of the quadrilateral through
opposite corners thus forming a parallelogram which is twice the area of the quadrilateral.
The parallelogram formed by the mid-points of the sides of the quadrilateral is clearly half the
dimensions of the surrounding parallelogram and therefore one-quarter of its area and half the
area of the quadrilateral.

THE REGULAR HEXAGON

This is a good figure to explore, for which the use of 1 cm equilateral triangle paper has been
found particularly helpful. The mid-point sequence of regular hexagons can soon be found to
have an area ratio for each stage of 0.75, either by counting triangles or by use of the mid-
point theorem.

    For quarter-points a starting hexagon of side 4 cm will result in the next three hexagons all
having their vertices conveniently on the grid lines of the paper. The area ratio is now 13/16.
With strategic choice of the size of starting hexagon some pupils have produced these results
for the regular hexagon. Clearly this pattern can be extended and generalized.

Corner position	$\frac{1}{2}$	$\frac{1}{3}$	$\frac{1}{4}$	$\frac{1}{5}$	$\frac{1}{6}$
Area ratio	$\frac{3}{4}$	$\frac{7}{9}$	$\frac{13}{16}$	$\frac{21}{25}$	$\frac{31}{36}$

THE REGULAR PENTAGON

This shape can be explored also but is not as tractable as the previous figures. There is the initial problem of drawing the first pentagon reasonably accurately, for which there are a variety of well-known methods. A challenge to demonstrate that the first mid-point pentagon is more than half the area of the original pentagon produces a response which involves folding the corners of the starting pentagon in along the sides of the second one, thus outlining the familiar pentagram inside the second pentagon. ('What has been cut off is less than half of what is left'.)

The area ratio can be linked to the famous Golden Section, which is the ratio of the side to the diagonal in a regular pentagon.

REVERSING THE SEQUENCE

This has been suggested in the following manner. With a square, each side is extended to double its length, working clockwise round the square. The ends of the extended sides are then joined up to form another square. It is of course rapidly remarked that this method, though consistent with the existing method, does not produce diagrams similar to the previous ones. However, sequences produced by the following two methods do produce rather similar diagrams:

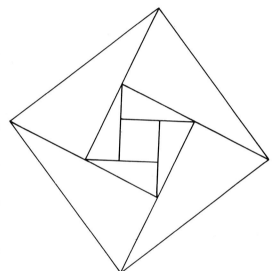

- internal points 0.1 along each side; and
- extending each side by 0.1 of its length.

Each new square in this diagram is found to be five times the area of its predecessor. Clearly this line of exploration can be extended.

EXTENSIONS

All the following ideas are for older pupils and students.

Limit point

The position of the limit point is perhaps intuitively clear for the regular figures and for parallelograms. However, for other figures, whether or not its position is dependent on the fraction chosen for marking the points, is worthy of consideration.

Hexagons: the general case

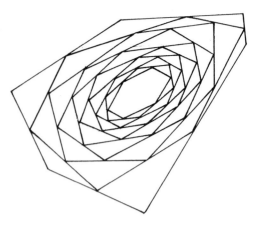

The quadrilateral case where a sequence of parallelograms is immediately created from the first stage onwards reveals an initial insight into generalizing regular pentagons and hexagons to irregular ones. The diagram here shows a mid-point sequence starting with an irregular hexagon.

The sequence soon settles down into two sets of alternate hexagons similarly orientated. Furthermore it can soon be observed that these hexagons have three pairs of approximately equal and parallel sides. This is a hexagonal analogue of the parallelogram sometimes called a **parhexagon**.

Experiment can now focus on parhexagons. Does a parhexagon start (with mid-points) straightaway produce another parhexagon, and does the third one in the sequence have its sides parallel to the initial parhexagon and so on? Squared paper is useful for creating parhexagons, for example

$(3, 0)$ $(0, 1)$ $(0, 3)$ $(5, 5)$ $(8, 4)$ $(8, 2)$

or

$(3, 0)$ $(0, 1)$ $(1, 3)$ $(5, 4)$ $(8, 3)$ $(7, 1)$.

Using vectors

Vectors are helpful for analysing the general cases of pentagons and hexagons. With pentagons the numbers of the **Fibonacci** sequence should soon be a feature, which is not surprising since that sequence has close links with the Golden Section.

The general problem has attracted the attention of academic mathematicians occasionally over the last forty years and it has been established that the limiting form of the figures is always a parallel projection or affine transformation of the corresponding regular figure. The following diagram shows a mid-point sequence of such pentagons.

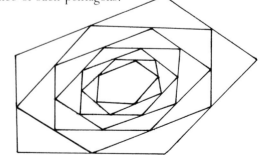

Though the two-dimensional case produces a stable convergent process, the three-dimensional case from which this investigation was first derived surprisingly produces a divergent process in which discrepancies from a regular form are rapidly magnified.

STATEMENT OF PROBLEM

Two lighthouses, A and B, each have a rotating light which sends a beam simultaneously in opposite directions as shown in this diagram. X is the spot illuminated by beams from both A and B.

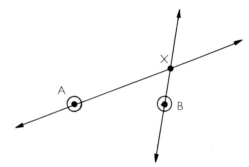

The lights flash on only at certain positions as they rotate. For example if they flash on only at each 45° turn from the starting position in the first diagram then the four points illuminated by both beams are shown in the diagram below. These would appear to be the corners of a square.

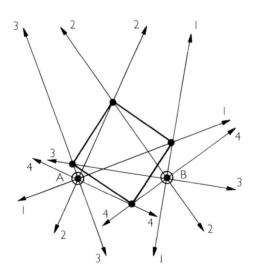

Draw a diagram in which the starting positions are different from the first diagram. Do the four points still form a square?

Under what conditions would only three points be illuminated?

Investigate the effects of varying the starting angles and flashing positions.

COMMENTARY

National Curriculum

Level 7 of AT11 refers to the **locus** of an object, moving subject to a rule, which is clearly the substance of this investigation. Conventional Cartesian coordinates are first mentioned earlier at level 4, while the coordinate system in use here is **angular bipolar** (as defined below) and has much in common with **bearings** (level 6).

The investigation is a suitable precursor to **angle properties of a circle**, the formal study of which is level 10 of AT10.

Materials

A circular protractor is useful for hand drawings to start with. However to reduce the work load involved in the production of diagrams, a specialized graph paper should be duplicated, consisting of two sets of radiating lines at equal intervals through two points placed centrally on the paper about 5 cm apart. The choice of the angle for the interval could be $15°$ which would give 12 lines through each point or $10°$ giving 18 lines through each point. One of each of these diagrams neatly fits on one side of an A4 sheet, so together they allow for an interesting range of variations of conditions. If too many lines are drawn in a diagram difficulty is experienced in distinguishing one beam from another.

GENERAL COMMENTS

The investigation was originally developed in relation to a particular geometrical result, the constant angle property of a circle. If the two beams are rotated in the same direction at the same angular speed but out of phase (that is, the beams are never parallel) then the illuminated points will lie on a circle through the two fixed points.

The principle of treating a mathematical result in an investigative manner has already been discussed in Chapter II, Sources of Investigations. It is certainly a possible approach but objectives can become confused; the investigative atmosphere can be thwarted if the desired result does not appear to emerge in the progress of the exploration and is forced as a consequence.

The investigation as framed can be very successful if normal investigative procedures and objectives are adhered to. Indeed experience with it can be related to the geometrical result which gave it birth, if that result is being considered subsequently. This seems a productive method of linking two types of mathematical experience.

A point located by *angular bipolar* coordinates is defined uniquely. The coordinates are the angles, measured positive anti-clockwise in degrees in the interval 0 to 180, which the lines joining the point to two fixed points make to a fixed direction. For convenience this direction can be the line joining the two fixed points. An ordered pair notation can be used, for example the starting position in the first diagram in the statement of the problem is (20, 80).

INITIAL APPROACHES

Some of the simpler drawings involving only a small number of flashes can be constructed with a circular protractor. First, draw the starting positions and then, using these beam lines as base lines, mark the positions of the beams for each flash.

Then the specialized graph paper referred to above should be used. Figures formed by joining up the illuminated points in order should be in a different colour from the grid lines. An example of such a diagram is shown below; there are three squares corresponding to the starting positions (45, 75), (30, 90) and (15, 105), with beams flashing at 45° intervals.

A square will always be produced if both lights flash on at 45° intervals no matter what the starting positions are, unless of course the beams are in phase (always parallel). The size of the square is dependent on the angular difference between the starting positions and pupils soon notice a relationship if they are encouraged to explore varying these positions. They can be asked to find what angular difference gives rise to the smallest square and which to the largest.

Precise explanations of why it is a square should not be expected. Angle properties of a circle, not surprisingly, do provide an explanation: each of the sides is a chord subtending an angle of 45° at the circumference and since equal chords subtend equal angles the four sides are equal, and also the angles at each vertex are 90°.

The reason why the three squares in the above diagram are all similarly orientated also derives from angle properties of a circle; all the squares have the sum of their starting positions 120°.

TABULATING BEFORE PLOTTING

To ensure correct plotting of points it is helpful to tabulate the points first, particularly if the step size is large. The set of four points in the second diagram in the statment of the problem is tabulated as follows.

	Flash interval	Start			
Beam from A	45	20	65	110	155
Beam from B	45	80	125	170	35

Note that fourth position of beam B is 35 and not 215, because we are using arithmetic modulo 180.

An algebraic description is formed by denoting the beam positions at A and B by x and y, and noting that all of the points satisfy the equation

$$y = (x + 60)(\mathrm{mod}\ 180).$$

POLYGONS AND STAR POLYGONS

Keeping the flash intervals equal but of a different size results in regular or star polygons. An example of the latter is tabulated as follows, giving a nine-pointed star.

	Flash interval	Start								
Beam from A	80	0	80	160	60	140	40	120	20	100	0
Beam from B	80	90	170	70	150	50	130	30	110	10	90

It is advisable to number the points whilst plotting, and then to join them up in order.

A flash interval of $20°$, however, gives a nine-sided regular polygon. Pupils should speculate what determines how many sides the polygon has and whether it is regular or star.

With smaller flash intervals plotting can be done by eye. A good diagram to draw is that produced by flash intervals of $10°$ with the results of a number of different starting positions superimposed.

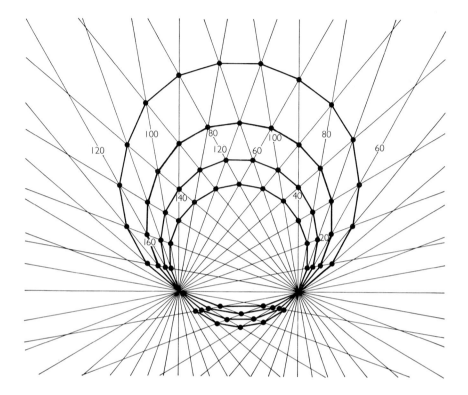

This gives a set of 18-sided polygons (approaching circles) all passing through the two base points as shown above. This is a very effective diagram for illustrating the constant angle property of a circle.

BEAMS ROTATING IN OPPOSITE DIRECTIONS

The direction of rotation of one beam can be changed while still retaining equal flash intervals, say of 10°. A simple case of this is where the beams both start on the zero line; this set of points has $x + y = 0 \pmod{180}$. A good diagram to draw includes this case and others based on the same rotation principle, for example

$x + y = 30 \pmod{180}$
$x + y = 60 \pmod{180}$
$x + y = 90 \pmod{180}$.

It may not be possible to plot some of the points when the two beams become nearly parallel.

DIFFERING SPEEDS OF ROTATION

Another simple variation is to keep the direction of rotation the same but to have one light flashing at double the interval of the other.

	Flash interval	Start								
Beam from A	20	0	20	40	60	80	100	120	140	160	0
Beam from B	40	0	40	80	120	160	20	60	100	140	0

Apart from the problem of how to deal with the start position in which the beams coincide (which should be discussed), the remaining part of the figure forms seven of the sides of a regular nine-sided polygon. It is not difficult, geometrically, to explain why all the points are distant AB from B. The points all satisfy $y = 2x$ (mod 180); a similar result is obtained for $x = 2y$ (mod 180).

Another case which follows the same principle but with different start conditions is $y = (2x + 90)(\text{mod } 180)$.

COMPUTER GRAPHICS

These are very effective in enhancing the exploration of the problem since a much wider range of cases can be examined which are otherwise inaccessible. With small flash intervals, curves rather than line figures are created. Programming, however, is not simple, the chief problem being with locating the point where the beams meet. More recent computing techniques using GEM software could ease the problem considerably.

EXTENSIONS

The investigation involves an exploration of an unfamiliar coordinate system. Other coordinate systems which have been similarly explored using specially prepared graph papers are listed below.

- **Bipolar**, using two sets of equally-spaced concentric circles like ripples spreading from two points; a family of figures in which $x + y = k$ for various values of k is a set of confocal ellipses with the two centres as foci.

- A combination of polar and Cartesian coordinates using a set of parallel lines and a set of concentric circles; points are located by their distance from a fixed point and from a fixed line.

INVESTIGATIONS: SHORT COMMENTARIES

19 GRACEFUL FIGURES

STATEMENT OF PROBLEM

In the diagram the vertices and sides of a triangle have been labelled with the integers 1 to 6 in such a way that the integer on each side is the positive difference between the integers at its two end vertices. Such a labelling is said to be *graceful* when the highest integer used is kept as small as possible and no integer is used twice.

Can you label the triangle gracefully in a different way?

Try labelling other figures gracefully.

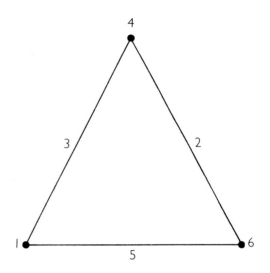

COMMENTARY

National Curriculum

The investigation requires only simple addition and subtraction (level 2, AT3), and recognition of 2-D figures (level 2, AT10). If figures other than triangles and quadrilaterals are to be explored then level 3 is more appropriate.

GENERAL COMMENTS

The originator of the term graceful in this context is thought to be S.W. Golomb in connection with a conjecture about labelling trees (see the investigation TREES (26)). A tree with n vertices has $n-1$ branches, and it is conjectured that all trees can be labelled *gracefully* using the first n integers for the vertices and the first $n-1$ integers for the branches. The statement of this investigation imposes a tighter constraint by requiring that the integers used for labelling are all different.

OTHER SOLUTIONS AND FIGURES

The triangle has another graceful solution, a quadrilateral appears also to have two solutions using the integers 1 to 8, and a pentagon, four solutions using the integers 1 to 10. Beyond allocating the largest number to a vertex good strategies for developing solutions have been hard to come by. Most though not all solutions have more of the high numbers allocated to vertices, for example $(1, 6, 10, 7, 9)$ for a pentagon.

At least eight distinct hexagon solutions have been found with two, curiously, using the same numbers for the vertices but in a different order, $(2, 10, 4, 7, 12, 11)$ and $(2, 7, 10, 4, 12, 11)$.

COMPLETE FIGURES

The triangle is a *complete* three point figure in that every point is joined to every other point. The corresponding figure with four points is shown. With six edges joining the vertices there is more constraint on solutions than with the quadrilateral case, and effectively the diagonals of the quadrilateral are now part of the graceful labelling scheme. No solution seems to exist using just the integers 1 to 10. However using ten of the first eleven integers pupils have found several distinct solutions and all, curiously, omit either 6 or 10.

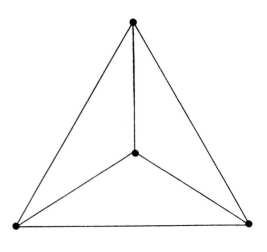

In a complete five-point figure there are ten edges (for example a pentagon with all its diagonals). Several solutions have been found with 17 as the highest integer but none with 16 or 15.

STATEMENT OF PROBLEM

What is the least number of coins that you need to move to change the triangle on the left into the triangle on the right?

Try it with different numbers of coins and possibly with other shapes.

COMMENTARY

National Curriculum

Recognizing triangles is level 2 (AT10) and ideas of **reflective symmetry** (level 3, AT11) are also involved. Level 3 is also more appropriate when larger triangles are explored and the sequence of solutions is developed.

Materials

Young pupils use coins or counters; older pupils often resort to using triangled paper.
   Squared paper is useful when right-angle turns are considered.

INITIAL APPROACHES

The temptation, at first, is to move all but the four base-line coins but alternative methods are soon spotted.
   For a 3-coin triangle, clearly one coin moved is sufficient and with a 6-coin triangle, two moves are sufficient.

Extending to 15-coin triangles and so on eventually gives rise to the results shown below. Note that the numbers in brackets refer to the coins to be moved at each corner of the triangle.

				Coins			
	1	3	6	10	15	21	28
Rows in triangle	1	2	3	4	5	6	7
Minimum moves for inversion	0	1	2	3	5	7	9
		(1)	(1+1)	(1+1+1)	(1+1+3)	(1+3+3)	(3+3+3)

AN OVERLAPPING METHOD

One novel approach to resolving the problem is to place the inverted triangle over the original one so that as many coins of the inverted triangle as possible are on top of existing coins, as illustrated in the diagram. The three non-overlapping coins have to be moved. This idea is then followed up by finding how many different ways inversion could be achieved. Clearly all ten coins could be moved (attempting no overlap). It is possible to move any number of coins from three to ten except for five, these being indicated by the impossibility of a 5-coin overlap. For a 15-coin triangle from five to

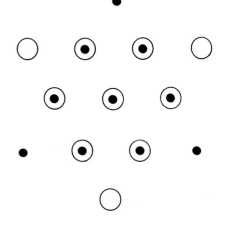

fifteen coins can be moved except for eight (no 7-coin overlap). How to predict which numbers of moves are impossible in larger triangles has not been resolved.

RIGHT-ANGLE TURNS

A triangle of coins can also be drawn on
squared paper. From the overlap shown in
this diagram it is clearly also possible to turn
a 6-coin triangle through 90° with a minimum move of two coins, which is the same as
required for inversion. Is this equally true for
other triangles and if so why?

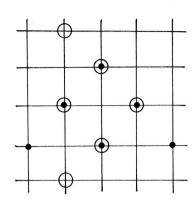

OTHER IDEAS

A simpler problem is to consider how many coins have to be moved to rotate a rectangular
arrangement through 90°. Older pupils should generalize this quite readily.

A generalization of the triangle problem is to examine inverting isosceles trapeziums of
coins (truncated triangles), and also turning them through 60°.

21 CHOP

STATEMENT OF PROBLEM

A pancake can be chopped into three or four pieces with two straight chops as shown.

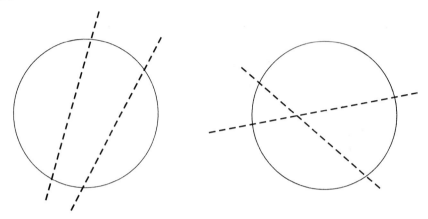

What is the smallest number of pieces into which a pancake can be divided by three straight chops (each chop *must* cut the pancake somewhere!).

What is the largest number of pieces?

What if four chops are made?

COMMENTARY

National Curriculum

Explaining **number patterns** is level 3 (AT5) and this is also the relevant level for the geometrical ideas involved (AT10).

The analogous three-dimensional problem is for level 5 upwards.

Materials

Young pupils could slice a layer of rolled plasticine. Drawing lines on plain paper is the obvious way to model the problem.

INITIAL APPROACHES

The minimum problem is simple and analogous to the corresponding one-dimensional problem of chopping a line, where n chops give $n+1$ pieces.

The largest number of pieces with three chops is seven. Pupils should comment on the differing nature of the pieces, one of the seven being completely inside the pancake, the others being bounded by either two or three chops and an edge of the pancake.

With a further chop the aim is to cut as many as possible of the existing pieces into two. Can the one-dimensional problem, already resolved, help here?

LOOKING AT THE PIECES

In the maximum case with three chops, there is one internal triangular piece. With four chops there will be three internal pieces; two triangles and one quadrilateral.

What is the distribution of internal triangle and quadrilateral pieces for five chops? Is it always the same, regardless of how the chops are made, providing of course the maximum number of pieces is obtained?

What is the maximum number of internal triangle pieces that can be achieved with six chops?

This development of the problem can be extended to include all pieces if the initial shape of the pancake is for example a triangle or a quadrilateral.

THREE DIMENSIONS

Chopping an apple or lump of plasticine by plane cuts is easy enough to handle and visualize up to three cuts. The maximum number of pieces follows the sequence 1 (no cuts), 2, 4, 8, . . .; what next?

It will soon be realized that all eight existing pieces cannot be chopped into two by a further cut. The two-dimensional problem already investigated may help in resolving this difficulty; a further cut is intersected by the existing cuts in a pattern of straight lines.

22   PALINDROMIC SUMS

STATEMENT OF PROBLEM

Choose a positive integer with two
digits, for example 67. Reverse the
digits and add the two numbers
together. Now reverse the digits of
the sum and add again. This time the
sum is a *palindromic* number, 484.

$$
\begin{array}{r}
67 \\
+\ 76 \\
\hline
143 \\
+\,341 \\
\hline
484
\end{array}
$$

Try again with a different starting number, with a different number of digits if you wish. How many steps does it take to get to a palindromic sum?

COMMENTARY

National Curriculum

Reasonable proficiency at addition is required, which suggests a minimum of about level 3 (AT3).

Materials

Use of a calculator is helpful both as a check and to speed up exploration. However eventually its limited digit capacity can be a serious shortcoming.

Squared paper is useful for tabulating observations.

GENERAL COMMENTS

This is a fairly well-known problem and usually attention has been focused on starting numbers such as 196 which apparently do not ever produce palindromic sums. The problem attracted the attention of Martin Gardner who wrote an article about it.[1] However many aspects of the problem remain unresolved and it has a continuing fascination.

TWO-DIGIT NUMBERS

Full analysis of all the two digits numbers is a good exercise in arithmetic. Alternatively a calculator can be very helpful until its digit capacity is exceeded, keeping a record of each sum up to the palindrome. It will soon be realized that integers with the same digital sum perform in the same manner. That all steps are multiples of 11 may also be noticed and this is reasonably explicable.

Results can be neatly tabulated in a double-entry table.

PROGRAMMING THE PROBLEM

The observation that 89 and 98 take 24 steps to reach palindromic state, whereas all other two digit integers take no more than 6 steps is quite strange. This has sometimes prompted a search for numbers which reach palindromic state in more than 24 steps. The greatest number of steps so far discovered is 29 (but see the later section on other bases). Hand calculations with this number of steps are prone to error and so a computer program is useful in which integers are stored and operated on as a list of digits. This is a good programming exercise for older pupils and students, and once written it is easily transferable to other bases. The chance of a palindrome clearly recedes as the number of steps increases. Some of the early steps in the process started by 196 are almost palindromes such as 887, 94039 and 897100798 but subsequently steps become more remote from palindromic form.

SEQUENCES OF STARTING NUMBERS

These can produce interesting results; for example 5, 55, 555, 5555, . . . all produce clearly related palindromes in two steps. The fact that these starting numbers are palindromes themselves does not preclude the routine being applied to them.

OTHER BASES

Working in other number systems is a good extension, with the help of a computer program when the starting numbers have more than three digits. Base 9 seems to provide few surprises with all two-digit numbers forming palindromes in at most seven steps. In base 8, starting with 777 produces 777777 after six steps and starting with 7777 produces a clearly identifiable asymmetric cycle every fourth step. Base 7 throws up some numbers with more steps to a palindrome than so far discovered in base 10, for example 56666 which takes 41 steps to become palindromic!

BINARY NUMBERS

Martin Gardner records that it has been established beyond doubt that some binary starting numbers *never* produce palindromes, the smallest being 10110 which produces a recurring asymmetric pattern every fourth step in a similar manner to the base 8 case already noted above. Clearly many more surprises are in store.

REFERENCES

1 Gardner, M. *Scientific American*, August 1970.

23   NUMBER SPIRAL

STATEMENT OF PROBLEM

On squared paper construct a number spiral as shown, continuing the spiral until at least 100.

Colour some of the numbers, for example even or square numbers, and try to explain the patterns formed.

Start a number spiral with 11 instead of 1 and colour the prime numbers.

Experiment with similar spirals on triangled paper.

17	16	15	14	13
18	5	4	3	12
	6	1	2	11
	7	8	9	10

COMMENTARY

National Curriculum

Distinguishing between odd and even is level 2 (AT5) and explaining **number patterns** is level 3. **Square** and **prime** numbers are referred to at level 5.

Materials

Squared paper is necessary for drawing the spiral and analogous spirals can be drawn on triangled paper.

PATTERNS OF ODD/EVEN AND SQUARE NUMBERS

The chessboard pattern of evens and odds is not directly connected with the method of constructing the spiral. Any method of filling the squares by inserting numbers consecutively which involves moving from a square to one of the four with which it shares a side will inevitably produce such a pattern. Some see this intuitively and a wide range of sensible verbal explanations can be expected.

Explanations of the diagonal arrangement of square numbers is perhaps easier. The 3-by-3 square around 1 is completed when 9 is reached and so on.

PRIME NUMBERS

When starting the spiral with 11 it is fascinating to see some of the primes forming a diagonal pattern. Will this continue as the spiral grows?

Why is 11 suggested as the starting number? Try some other starting numbers.

TRIANGULAR SPIRALS

With triangled paper the odd/even arrangement is often referred to as odd 'point up', even 'point down', or vice versa, and the explanation of this is similar to that already mentioned for squared paper.

It seems natural to shade the triangle numbers on triangled paper and if the spiral is taken far enough (to at least 500) then a seven-curved spoke arrangement of the triangle numbers emerges. Can this be explained, if only partially?

24 CHESSBOARD RECTANGLES

STATEMENT OF PROBLEM

The 4-by-5 'chessboard' shown here has been shaded randomly with ten black and ten white squares. Unfortunately four of the black squares, A, B, C and D are at the corners of a rectangle. Can you rearrange the shading so that no four black or four white squares are at the corners of a rectangle or square?

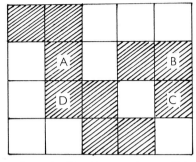

Experiment with smaller and larger chessboards. Do the numbers of black and white squares have to be equal?

COMMENTARY

National Curriculum

Level 3 with its references to sorting 2-D shapes (AT10) and **reflective symmetry** (AT11) would appear to be the minimum.

The arguments used in resolving the largest possible board which meets the requirements need combinatorial skills at about level 5 or 6.

Materials

Squared paper is obviously very useful.

Young pupils can create squares and rectangles using two different colours of cube.

GENERAL COMMENTS

Making square D white and the one to its left black fulfils the requirements about the corners of subsidiary rectangles and squares. If the words *or square* are omitted from the requirements some pupils assume that an arrangement of four similarly coloured squares at the corners of a larger square is allowable, but not at the corners of a rectangle. (Are squares rectangles?) This relaxes the constraints a little and the investigation can be developed accordingly if so wished. Otherwise, this approach could be treated as an extension.

Lettering the squares, for example O and X, is a good and speedier alternative to colouring for older pupils.

INITIAL APPROACHES

After initial experimentation with the given figure it is wise to start by concentrating on smaller boards. With a 2-by-2 board some pupils look at the number of different ways the colouring can be achieved, the only inadmissible cases being all white and all black. Binary notation has been used with all the possible colourings corresponding to the binary numbers from 0001 to 1110 inclusive. Geometrically many of these fourteen cases are equivalent by reflection or rotation, reducing the number to four. It is also often remarked that to each solution there is an inverse with white and black interchanged.

BOARDS OF WIDTH 2

With a 2-by-3 board it is rapidly realized that there must be at least two black or two white squares. Can these be randomly placed on the board? Enumeration of cases is no longer a simple matter, but the binary representation would seem to suggest that 44 of the binary numbers from 000000 to 111111 correspond to acceptable arrangements.

The observations, though not necessarily the enumeration, can be generalized to 2-by-n boards for which successful colourings are always seen to be possible.

BOARDS OF WIDTH 3 OR MORE

Moving to the next size of board, 3-by-3, finding a successful arrangement using only three black (or white) squares comes as a surprise. From this point onwards progress with the investigation is usually frustrated until some system is introduced into the colouring process. A sudden insight often occurs that columns (or rows) cannot be repeated. Successful 3-by-4, 3-by-5 boards and so on are then drawn by adding different columns to the already successful 3-by-3 board. In particular, adding a column with all black squares to the 3-by-3 board (or all white if the 3-by-3 board has 3 white and 6 black squares) is found to inhibit further expansion.

There is thus clearly a limit to the length of board of width 3 for colouring to be successful. What then of boards of width 4? . . . of width 5?

EXTENSIONS

Two extensions of the investigation are regularly suggested. To consider cuboids in three dimensions is clearly quite a challenge for the older pupil, and a more tractable problem is to use three colours in the present two-dimensional context. Perhaps in this case a successful colouring of a 3-by-n board is always possible. What are the limits, if any, on a 4-by-n board?

25 SPIROLATERALS

STATEMENT OF PROBLEM

On centimetre-squared paper follow the instructions in the flow chart shown here, drawing the path taken as you go. You should find yourself going round and round a rectangle 2 cm by 3 cm. We could call this path a spirolateral.

Using T to mean 'turn clockwise' and M to mean 'move forward', mark a starting point and then draw the path of the following spirolateral.

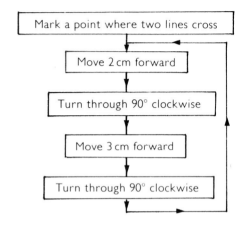

Did you get back to your starting position? If so how many moves did it take?
Invent and draw some other spirolaterals and try to explain your observations.

COMMENTARY

National Curriculum

Level 2 (AT11) refers to understanding instructions for turning through right angles, which is clearly the basic requirement for the problem. Use of degrees is not necessary until the later development when the angle is altered. At this level referring to movement in centimetres should be replaced by paces, or squares if squared paper is used.

Materials

Physical enactment of instructions is best done on a square-tiled surfacing. Otherwise recording should be done on squared paper, or triangled paper for the 120° and 60° variations.

GENERAL COMMENTS

The original source for the formulation of this investigation is due to D.R. Byrkit (1971)[1], and another article in the same journal two years later by F.C. Odds[2] used the term spirolaterals in a similar context.

Several computer programs are available commercially which can be very helpful in the exploration of some spirolaterals. Use of BASIC graphics or LOGO is straightforward.

The investigation VECTOR PATHS (36) has distinct structural affinities with the present one, although it uses a different notation and is more general.

INITIAL APPROACHES

The second spirolateral in the statement of the problem returns to the starting point after four *cycles*. Having three specific moves in each cycle we could say that it has *order* 3. The first spirolateral has order 2 and is complete after 2 cycles.

For initial experimentation it could be wise to keep to 90° clockwise turns and vary the lengths and number of moves in each cycle. It has also proved a good idea to clearly mark the position at the end of each cycle. The diagram shows a spirolateral which is of order 6 and is complete after two cycles. Once a good number of cases have been drawn a relationship between the order and number of cycles should be suggested. Some spirolaterals are found, perhaps surprisingly, never to return home.

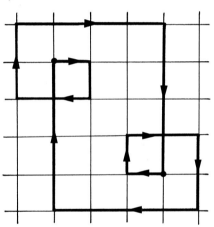

HOW MANY CYCLES?

All odd-ordered spirolaterals (with 90° clockwise turns) have 4 cycles. Those whose order is a multiple of 4 are infinite except when one cycle is itself complete; the other even-ordered ones have two cycles. How is this all to be explained?

The simplest spirolateral with one move and one 90° turn in each cycle will clearly traverse the four sides of a square. For the order-three spirolateral in the statement of the problem, join the starting point to the position at the end of the first cycle, then join this position to the position at the end of the second cycle and so on. Does this help?

OTHER TYPES OF SPIROLATERAL

Possible variations to the definition of a spirolateral which have been explored are:

- altering the angle of turn;
- involving two or more different angles in a cycle;
- introducing backwards moves.

Where the same angle is used throughout, a spirolateral, a simple notation can be used. The order-three spirolateral in the statement of the problem has three movement of 1, 2 and 3 with the same 90° angle between each move, and hence could be denoted (90°, 1, 2, 3). In this notation (120°, 1) clearly gives an equilateral triangle.

What would give a regular hexagon? (We have now turned the problem on its head by starting with a figure and designing a spirolateral to produce it.) One answer is fairly obvious but an interesting one which was suggested is (120°, 1, ⁻1). Are there other possibilities? What does (90°, 1, ⁻1, 1) give?

REFERENCES

1 Byrkit, D.R. (1971) Taxicab geometry, *Mathematics Teacher*, NCTM, May.
2 Odds, F.C. (1973) Spirolaterals, *Mathematics Teacher*, February, 121–4.

26   TREES

STATEMENT OF PROBLEM

A tree is drawn by connecting points
by branches (lines) without forming
any closed loops. An example of a
tree with eight points is shown in this
diagram.

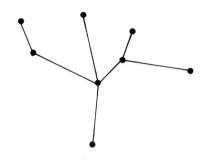

Investigate the number of different
trees which can be drawn using 1, 2,
3, 4, . . . points.

COMMENTARY

National Curriculum

Level 5 of AT11 refers to drawing **networks** to solve problems. However the rudimentary
use of networks in this problem would seem likely only to require level 3.

If systematic enumeration of trees with more than six points is tackled then level 4 or 5 is
more appropriate.

GENERAL COMMENTS

Trees are a subset of graphs or networks. The original source for this investigation is Ore[1]
which together with Wilson[2] provide further ideas for similar investigations.

A few initial problems often need to be clarified.

- For a chosen number of points, a path must be possible along branches from any point to
  any other point of the tree. Otherwise there is more than one tree. On the other hand, a
  diagram containing more than one tree could be called a 'forest'. The number of forests
  that can be drawn using 6 points say, clearly contains the 6-point trees as a subset unless
  it is specified that a forest must contain at least two trees!

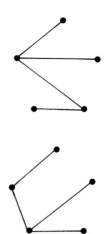

- What constitutes different in respect of two trees with the same number of points needs resolution. If 5 points are placed in the corner positions of a regular pentagon, as pupils frequently do, then are the two trees in this diagram different or not? The idea of **topological equivalence** is relevant here and quite young pupils understand it once they envisage the trees as flexible wire frameworks. (See comments on ATs 10 and 11 in Chapter V.)

POINTS AND BRANCHES

A simple relation between the number of points and branches in a tree is frequently noticed after an initial period of experimentation, and explanation or proof of this at different levels of sophistication are forthcoming. The result can be extended to include forests.

CLASSIFYING TREES

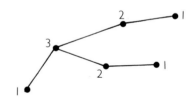

A typical method of attempting to classify topologically different trees is to label the points with the number of branches emanating from them as shown in the diagram. This six-point tree could then be referred to as $(3, 2, 2, 1, 1, 1)$ with a convention of listing the numbers in descending order.

This notation gives rise to a number of questions which require some sort of resolution.

- Why are at least two of the numbers always 1?
- Is the sum of the numbers always the same for a given number of points?
- Is a tree *uniquely* described by an ordered set of numbers? For example is it possible to draw a topologically different tree with the set $(3, 2, 2, 1, 1, 1)$? If so, which sets of numbers uniquely define trees and which do not?

Once such questions are resolved, it is possible to make some progress towards finding the number of trees for a given number of points. For example it is not difficult to enumerate for six-point trees the 5 sets $(5, 1, 1, 1, 1, 1)$, $(4, 2, 1, 1, 1, 1)$, $(3, 3, 1, 1, 1, 1)$, $(3, 2, 2, 1, 1, 1)$, and $(2, 2, 2, 2, 1, 1)$ where the numbers in each set total 10. It will then be found that six different trees can be drawn, one of the sets giving rise to two distinct trees.

TREES WITH MORE THAN SIX POINTS

For seven-point trees there are seven sets of numbers and for eight-point trees 11 sets, with the tree in the statement of the problem corresponding to $(4, 3, 2, 1, 1, 1, 1, 1)$. How many other trees can be drawn with this set? Using seven points it is conjectured that there are 11 possible trees and for eight points, 22 trees. The set $(3, 3, 2, 2, 1, 1, 1, 1)$ appears to give rise to the most trees with eight points.

Both sequences, of number sets and of trees, appear very difficult to generalize.

REFERENCES

1 Ore, O. (1963) *Graphs and Their Uses*, Random House, 34ff.
2 Wilson, R.J. (1972) *Introduction to Graph Theory*, Longman.

27 PAVING WITH DOMINOES

STATEMENT OF PROBLEM

In the diagram the patio (5 m by 5 m) surrounding a pond (1 m by 3 m) has been successfully paved with large 'domino' paving stones measuring 2 m by 1 m.

Suppose the pond is placed centrally, that is, 1 m down from its present position. Can the patio still be paved in the same manner?

Investigate paving other sizes and shapes of patio with or without ponds.

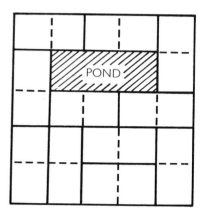

COMMENTARY

National Curriculum

Level 2 of AT10 refers to recognizing squares and rectangles, and creating patterns with 2-D shapes. Young pupils can experiment with dominoes and produce successful pavings. Explanations based on alternate colouring appear to require level 4 upwards.

Materials

Squared paper is invaluable at all stages.

For 3-D problems, number rods or their equivalent are essential, at least to start with, at all levels.

GENERAL COMMENTS

The context of this investigation is similar to that of WALLS (5) and CRAZY PAVING (14), but the problem is different.

It is a widely known result that a standard chessboard with two opposite corner squares deleted cannot be covered with 2-by-1 dominoes of the appropriate size. The clue lies in the colouring of the squares with each domino covering one of each colour; the two opposite corners are of the same colour and hence the remaining squares cannot be covered.

Colouring the squares of the patio in the statement of the problem alternately in chessboard fashion shows that the second case mentioned is impossible.

A not so well-known result is that an odd-dimensioned rectangular patio can be paved after any three squares are deleted such that equal numbers of squares of each colour remain. Paving is no longer always possible if five squares are similarly deleted, nor if two of each colour are deleted from a patio with at least one even dimension.

COLOURING ALTERNATE SQUARES

Having equal numbers of squares of each colour is clearly a necessary though not sufficient condition for paving with dominoes. The issue of whether or not to suggest alternate colouring of squares to a pupil or student exploring the problem is crucial and there is no easy answer to this dilemma.

Systematic examination of simple cases has however been found to lead to it. For example it is soon found that a 3-by-3 patio can be paved if a 1-by-1 pond occupies either a corner or the centre position but not if it is placed in the middle of a side. Colouring the impossible positions reveals the chessboard pattern. Similarly, a 5-by-3 patio with a 1-by-1 pond can be examined systematically.

ENUMERATING POSSIBLE PAVINGS

Where paving is possible, enumerating the number of distinct cases is a good development and not at all difficult in simple cases. The 3-by-3 with a missing corner has four distinct pavings in two mirror-image pairs.

LARGER PAVING STONES

An extension to triomino paving stones, either 3-by-1 or L shaped, is well worth pursuing. For the latter case, an interesting patio to start with is a 4-by-4 with one deleted square. Only three distinct cases arise each of which can be paved uniquely employing the result that a double-size L shape can be built from four L-shaped triominoes (see the investigation CLONES (15)).

THREE DIMENSIONS

A three-dimensional extension is to attempt to build various sizes of cuboid, with or without some missing cubes, with Cuisenaire rods of length 2.

28   LOOPS

STATEMENT OF PROBLEM

A loop of string 9 cm long can be
formed into an equilateral triangle
with side length 3 cm. The triangle in
this diagram with sides of length
2 cm, 3 cm and 4 cm can also be
formed from the loop.

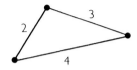

   Both triangles have a perimeter of
9 cm. Can you find another triangle with integer sides which can be formed from
the loop?

   Find sets of triangles formed from other lengths of loop.

   How many different triangles with integer sides do you think you could find with
a perimeter of 100 cm?

COMMENTARY

National Curriculum

It has already been remarked in the commentary on CUSTARD PIES (11) that the **triangle
inequality** (any two sides are together greater than the third) is probably level 4 or 5.

   The basic required number knowledge is addition of whole numbers, and to distinguish
between odd and even; this is level 2 or 3 (AT5).

   If triangles are to be constructed then level 4 (AT10) is necessary.

Materials

Finely accurate construction is not necessary, so use of a ruler and plain paper should suffice.

INITIAL APPROACHES

This problem looks quite straightforward initially. First reactions are frequently to take a
number and split it into three integers as many ways as possible; for example 7 is

$5+1+1$ or $4+2+1$ or $3+3+1$ or $3+2+2$,

hence suggesting four possible triangles with a perimeter of 7.

The best response to this situation is to suggest that the supposed triangles should be drawn with reasonable accuracy. The triangle inequality comes into play and the number of triangles is found to be only two, (3, 3, 1) and (3, 2, 2), using a popular triple notation. The problem is *not* quite so straightforward.

A very useful observation which is often made at this stage is that the longest side has to be less than half the perimeter. This is sufficient to ensure that triples with a given perimeter constructed with this condition are all valid triangles; for example, with perimeter 15 there are seven such triangles:

$$(7,7,1) \quad (7,6,2) \quad (7,5,3) \quad (7, 4, 4) \quad (6,6,3) \quad (6,5,4) \quad (5,5,5)$$

A GOOD SUPPLY OF DATA

Systematic exploration of perimeters from 3 upwards is now sensible and needs to be taken up to at least 27 if, eventually, predictions are to be made from an apparently erratic set of results.

						Perimeter							
	3	4	5	6	7	8	9	10	11	12	13	14	15
Triangles	1	0	1	1	2	1	3	2	4	3	5	4	7
	16	17	18	19	20	21	22	23	24	25	26	27	
Triangles	5	8	7	10	8	12	10	14	12	16	14	19	

It seems remarkable that a perimeter of 8 can produce only one triangle and that an increase in perimeter often results in a decrease in the number of possible triangles. Such is the fascination of the problem.

ALTERNATIVE FORMS OF DATA

All triangles with a perimeter of 8 or less are isosceles, and this leads some to look at the isosceles subset of all the triangles. Others may wish to distinguish, say, between $(2, 3, 4)$ and $(2, 4, 3)$ with perimeter 9, one being the mirror image of the other if the sides are listed in the triple conventionally in clockwise order. This increases the number of possible triangles, those with perimeter 15, for example, increasing to ten.

ODD AND EVEN PERIMETERS

Splitting the list into odd and even perimeters is a useful strategy and produces almost instant order out of chaos, since the two sequences are identical but out of phase by 3. This clearly calls for explanation. Why should a perimeter of 18, for example, give rise to the same number of triangles as a perimeter of 15?

Listing those with perimeter 18 and comparing them with the list of seven triangles for a perimeter of 15 listed earlier,

$$(8, 8, 2) \quad (8, 7, 3) \quad (8, 6, 4) \quad (8, 5, 5) \quad (7, 7, 4) \quad (7, 6, 5) \quad (6, 6, 6)$$

reveals a one-to-one correspondence, those with perimeter 18 having their sides each 1 longer. If 1 is now added to each side of the triangles with perimeter 18 then triangles with perimeter 21 are formed, but this time there are more than seven. Five of the twelve triangles with perimeter 21 cannot be obtained in this manner. Which five? Does this lead to an explanation of the correspondence?

GENERALIZING THE SEQUENCE

The differences between the numbers in the sequence for odd perimeters does begin to show a pattern, $0, 1, 1, 1, 1, 2$ followed by $1, 2, 2, 2, 2, 3$. The next six differences are plausibly $2, 3, 3, 3, 3, 4$ and results determined from this hypothesis can be checked. Some explanation however is called for as to why the difference pattern develops in sets of six for the odd perimeters and therefore in sets of twelve for all perimeters.

Given that the pattern of differences continues in this way then the number of triangles with a perimeter of 100 can be predicted.

29   MAGIC

STATEMENT OF PROBLEM

The 3-by-3 magic square of numbers in which all the rows, columns and diagonals add up to the same number will probably be familiar.

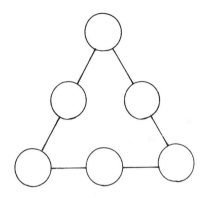

Instead, try to fit the numbers 1 to 6 into the magic triangle in the diagram making all the sides add up to the same number.

The magic square has three numbers to a side and one in the centre. Try to design a magic pentagon, a magic hexagon and so on.

COMMENTARY

National Curriculum

Addition of numbers at level 3 (AT5) is sufficient to start this investigation.

Complete exploration of the hexagon is probably possible at level 5 and arguments about impossibility, using algebra, require level 7 or 8.

Materials

The diagrams onto which the numbers are to be written could be duplicated.

GENERAL COMMENTS

Magic squares have a long history dating back to Chinese mathematics in the 3rd century BC. The 3-by-3 square is unique, apart from rotations and reflections, but the 4-by-4 has 880 distinct forms. Further development can be found in a chapter devoted to magic squares in a classic book by W.W. Rouse Ball[1] first published in 1892 and revised and reprinted on numerous occasions since.

MAGIC TRIANGLES

The magic triangle in the statement of the problem does not present much difficulty to a structured search. There are four distinct solutions with the sides adding up to 9, 10, 11 and 12. A magic triangle which is truly analogous to the 3-by-3 magic square should have a central number as shown here. However it should soon be discovered that a triangle cannot be completed with the numbers 1 to 7 such that the sums along all the lines are the same. What numbers can be inserted to make it have this magic property?

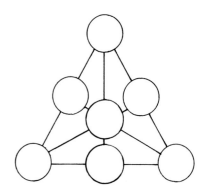

MAGIC PENTAGONS

A magic pentagon would need the numbers 1 to 11, with five at the vertices, five at the middles of the sides and one in the centre. The sums along the sides and along the lines joining each vertex to the middle to the opposite side would have to be equal. Symmetry suggests that, as with the 3-by-3 magic square, the middle number (6 in this case) should be in the centre spot. Perhaps surprisingly no solution has been found and subsequently explanation of this was formulated.

MAGIC HEXAGONS

A magic hexagon does have solutions, one being shown in this diagram. Four distinct ones have been found and these are interestingly related. All have 7 placed centrally and a sum of 21, both of which can be reasonably surmised and subsequently explained. The other numbers are associated in triples, for example (1, 8, 12) appears in the diagram along the top side with 8 in the middle. In other solutions it appears either along a side with 1 or 12 in the middle, or with the three numbers at the middle positions of alternate sides. This observation applies equally to 7 other triples. There is thus an additional magic property: not only do the sides and cross lines sum to 21 but also the middles of alternate sides.

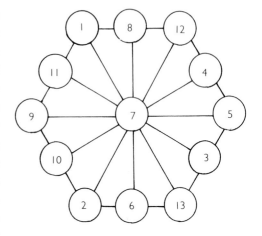

Whether or not magic heptagons exist has not been explored but magic octagons have been found.

MAGIC STAR POLYGONS

These are also worth exploring. For the hexagram the result in this diagram was found. Not only do the six quadruples along the sides sum to 26 but also the four corners of three rhombuses, for example $(1, 7, 6, 12)$. The inner hexagon also has the magic sum and the outer hexagon twice the sum.

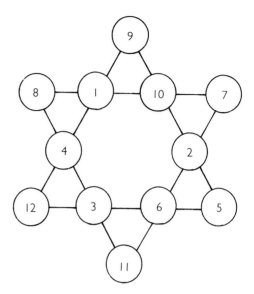

USING PRIME NUMBERS

In place of consecutive integers, Rouse Ball[1] gives a solution for a 3-by-3 magic square using only prime numbers all less than 100, his solution including 1 which is not strictly prime. If 1 is excluded then solutions are still possible but the smallest involves some primes greater than 100. Finding such prime magic squares requires a thoroughly structured approach and is best suited for advanced and higher level students. Younger pupils can however successfully tackle the problem of finding prime magic triangles in the form of the diagram in the statement of the problem.

REFERENCES

1 Rouse Ball, W.W. (1938, 11th edn) *Mathematical Recreations and Essays*, Macmillan, 193 ff.

STATEMENT OF PROBLEM

Eight centimetre cubes can be linked into a cube twice the size as shown. How many square centimetres are on the outside surface of this larger cube?

Now rearrange the cubes and link all eight together face to face in another way. What is its surface area?

Try to make shapes with different surface areas. What shape has the greatest surface area?

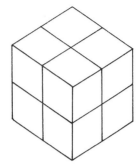

COMMENTARY

National Curriculum

Level 4 of AT9 has an example directly related to this investigation. The subject matter being **area**, **volume** and 3-D figures clearly identify it with AT8 and AT10, also at level 4.

2-D drawing of the 3-D shapes on squared or isometric paper comes later at level 6, which is also the more appropriate level for systematic exploration of the gap-filling aspect of the problem.

Materials

A sufficient supply of centicubes (of the commercially available linking variety which adhere face to face) is required for each pupil involved to have at least eight.

For recording purposes all pupils should make some attempt at drawing the shapes initially on plain paper. Older pupils can successfully use squared or triangled (isometric) paper, though care is needed when some cubes are hidden by others.

GENERAL COMMENTS

An exploratory approach has been used by teachers for many years to illuminate the independence of surface area and volume and, analogously in two dimensions, the independence of area and perimeter. This investigation is one approach to the former by considering a fixed volume in a variety of forms. There are numerous ways in which the eight centimetre cubes can be joined face to face, the extremes being a $2 \times 2 \times 2$ cube and a $1 \times 1 \times 8$ cuboid with surface areas of 24 cm² and 34 cm², respectively.

Instead of extending a sequence of numbers this investigation is an example of gap-filling. Can shapes be formed with surface areas equal to each of the values between the two extremes? If not, then why not?

INITIAL APPROACHES

Most pupils have little difficulty finding the 'line' case of maximum surface area. They should then be challenged to find intermediate ones, and some pupils may need encouragement to construct irregular shapes.

The first attempts at determining surface area will probably be the unsystematic counting of faces. This is prone to error, particularly with the many asymmetrical shapes which can be constructed.

When continual disagreements arise it is soon realized that a more systematic method is required. A frequently-used system is to count squares seen from top/bottom (plan), front/back and two side views (elevations). These views (or projections) can quite easily be drawn on squared paper by standing the shape on the paper and effectively drawing round what can be seen looking down from above the shape.

FILLING ALL THE GAPS

Initially it will not be realized that the number of faces visible from the top is the same as from the bottom and so on. Consequently odd numbers may be suggested for the surface area. Eventually, it is hypothesized that only even surface areas can occur.

Various sound geometrical justifications for this are likely to be put forward beside the counting system already considered, for example each linking of two cubes together renders two square faces invisible.

Can shapes be found with surface areas of 26, 28, 30 and 32? A surface area of 26 appears to be geometrically impossible (why?) and so the neat arithmetical filling of all the even numbers in the sequence cannot be achieved. Nevertheless a complete set of possible values satisfying the given requirements has been found and this itself is significant.

EXTENSIONS

Changing the number of cubes is the obvious extension. With nine linked cubes, are all even surface areas from 28 to 38 inclusive possible?

Other two- or three-dimensional figures can be similarly linked together and complete sets of possible perimeters or surface areas determined.

A related problem associated with drawing the plan and two elevations of a shape is to investigate whether the three such projections of the shape are unique to it. That is, given the plan, front and side elevations of a shape, is it uniquely determined?

31   DIFFERENCES

STATEMENT OF PROBLEM

Take any four-digit number, say 3761, and rearrange the digits first in descending order and then in ascending order. Subtract the second arrangement from the first. Repeat this process on the resulting four-digit number 6264 and so on.

$$\begin{array}{r} 7631 \\ -1367 \\ \hline 6264 \end{array}$$

Try the process on numbers with different numbers of digits, and perhaps in other bases.

COMMENTARY

National Curriculum

The arithmetic involved, which to some extent is self-checking due to the nature of the problem, is about level 4 (AT3).

Other bases receive no mention in the National Curriculum, and this is surprising as their counter example effect can strengthen understanding of the conventional base 10 routines.

Materials

Use of a calculator can speed up exploration.

GENERAL COMMENTS

The numbers 0000, 1111, 2222, . . ., 9999 clearly all result in 0000 after one subtraction and are a special but trivial case. When the result of a subtraction has less digits than the starting length, it should be preceded with enough initial zeros to bring it back to the correct length.

That all 4-digit numbers, except the trivial cases already noted, eventually result in 6174 (called **Kaprekar's constant**) may be known. Nevertheless the patterns in the way numbers converge on this constant are worthy of investigation.

A computer program is possible, in which the digits are stored separately in a list and a procedure for sorting them in order of size is required.

TWO-DIGIT NUMBERS

To start, it simplifies the problem to consider two-digit numbers. This is more obviously containable and it will soon be seen that after one reversal all numbers with two different digits give a number whose digital sum is 9. This observation can be generalized by older pupils.

All numbers thus link at some point into the continuing chain

$$09 \to 81 \to 63 \to 27 \to 45 \to 09.$$

This much simpler problem can help considerably in structuring the more complex cases with more digits.

THREE-DIGIT NUMBERS

With three-digit numbers, if the digits are not all the same there is an end-stop of 495 and again the pattern of the convergence can be analysed. The central 9 appears at the first subtraction and pupils should be able to explain this.

FOUR DIGITS AND MORE

At first sight it might seem a daunting task to cover all possible four-digit numbers but many pupils soon realize that any of the 24 arrangements of 1, 3, 6 and 7 will perform in precisely the same way as the example in the statement of the problem. Examining the relation between the four digits of the numbers which occur after the first subtraction will further reduce the size of the overall task significantly.

Generalizing the first subtraction will show that the resulting number has either

1. first and fourth digits totalling 9, and second and third digits both 9, or
2. first and fourth totalling 10, and second and third totalling 8.

For (1) there are only five possibilities and for (2) there are 25, so the full analysis of the 4-digit problem is now within reach.

With five digits, as with three digits, the central 9 is again a feature. Some numbers appear to converge into chains. One of these contains Kaprekar's constant 6174 with a 9 inserted centrally, which is intriguing if not surprising.

OTHER BASES

In other bases, for the four-digit case (with digits not all the same) various combinations of end-stops and/or chains will be found; those with just end-stops are base 2 (end-stop 0111) and base 5 (end-stop 3032). Is there any reason why the three bases which feature just end-stops are 2, 5 and 10?

Other digit cases can be examined similarly.

32 DUOGONS

STATEMENT OF PROBLEM

A duogon is a figure with two lines, each line starting and finishing at a point. The diagram shows three examples of duogons each with a different number of points.

    Can you find any more duogons?

COMMENTARY

National Curriculum

Networks are mentioned at level 5 or AT11, but their basic use here is earlier, at about level 3. Systematic enumeration of trigons is more likely to be level 5.

Materials

Drawings are best on plain paper.

GENERAL COMMENTS

This investigation belongs to the same area of mathematics as the investigation TREES (26), that of graph or network theory. The idea of the **topological equivalence** of diagrams is fundamental (see comments on ATs 10 and 11 in Chapter V). Unlike trees, duogons do not necessarily have to be *connected*, so a duogon can consist of two separate parts as for example, one of the three duogons in the statement of the problem.

INITIAL DISCUSSION

Some discussion is bound to occur on what constitute different duogons. Is the duogon in this diagram the same as one of the three in the statement of the problem or not? Envisaging them as flexible wire frameworks helps.

A SUITABLE NOTATION

Describing duogons by sets of numbers, conventionally arranged in ascending or descending order, can be helpful. For example, the three duogons in the statement of the problem could be described as $(2, 2)$, $(2, 1, 1)$ and (4) with each number corresponding to a point in the duogon.

As happens with a similar notation for trees, it will be found that distinct duogons can have the same set description. The notation however can help in the search for other duogons. Since the total of the numbers in the set has to be 4 (why?) duogons corresponding to $(3, 1)$ and $(1, 1, 1, 1)$ should be possible.

TRIGONS AND BEYOND

Trigons, figures with three lines and from one to six points (why no more than six?), are an obvious analogy to duogons. There are 23 topologically different trigons which have been found, though this number is dependent on the decision about what constitute different trigons.

The number of quadrogons is clearly going to be considerably larger.

Some early generalizations, for example about the number of n-gons with either 1, $n-1$, or n points, are often proposed.

33 BALANCING ACT

STATEMENT OF PROBLEM

You have a good supply of identical bricks (number rods or dominoes will do). Try to balance the bricks on the edge of a level table so that your pile overhangs as far as possible without overbalancing.

Suppose a brick is 2 cm long. Then a single brick can overhang by 1 cm and two bricks could overhang 1.5 cm as shown in the diagram, with brick A just balancing on brick B and the two together symmetrically placed with regard to the edge of the table.

Experiment with two or more bricks.

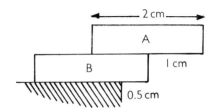

COMMENTARY

National Curriculum

This problem might be regarded strictly as part of the science curriculum, since mechanics has for many years been on the borderline between mathematics and physics.

Balancing fascinates children from a very early age, but this problem of moments involves reasoning about equilibrium and the mathematics involved is that of **ratio** and **proportion**, at level 5 or 6 of AT2.

Materials

Dominoes and number rods mentioned in the statement of the problem are good for the experimental aspects.

GENERAL COMMENTS

For a single stepped pile of bricks, extending the diagram in the statement of the problem to three or more bricks, the theoretical problem has the fairly well-known solution of

$$1 + \tfrac{1}{2} + \tfrac{1}{3} + \tfrac{1}{4} + \tfrac{1}{5} + \cdots$$

where each term of the series is the overhang of each brick over the one immediately underneath, starting from the top downwards. Since the sum of this series has no limit, in theory, any desired overhang can be achieved given sufficient bricks, which seems rather remarkable.

Two short notes in the *Mathematical Gazette*[1] discuss more subtle approaches which achieve greater overhangs. In the second R.E. Scraton points out that the maximum overhang which can be achieved by one brick is half the length of the longest face diagonal, not just half the length of the longest edge!

INITIAL APPROACHES

Experimental work gives a feel for practical constraints, for example slight imperfections in the bricks, and leads to discussion about these constraints.

Disregarding the quite complex solutions for four and more bricks mentioned in the previous section, there is plenty of scope for pupils and students to tackle the problem at their own level and produce interesting results.

This diagram is a suggested alternative to the one in the statement of the problem for two bricks and the diagram in the next section is thought to be the best solution for three bricks, certainly better than the single stepped pile overhang of $1\frac{5}{6}$.

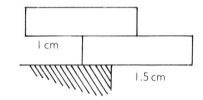

BALANCING FOUR BRICKS

If the bricks in this diagram are balanced on the edge of a fourth brick then that brick can have a small overhang.

A better solution can be produced by placing the fourth brick symmetrically on the top of these three bricks and then moving

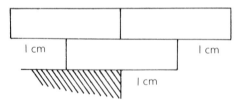

the two central bricks outwards away from each other keeping the whole edifice symmetrical.

Some pupils may know little or nothing about the theory of moments but nevertheless can produce valid arguments about counterbalancing to justify four-brick examples such as these, possibly based on practical experience of see-saws.

REFERENCES

1 Scraton, R.E. *Mathematical Gazette*, December 1979 and October 1980.

34 QUADRUPLES

STATEMENT OF PROBLEM

The triple $(3, 4, 5)$ is very well-known because $3^2 + 4^2 = 5^2$. It is sometimes called a **Pythagorean triple** because of Pythagoras' famous theorem about the sides of a right-angled triangle.

The quadruple $(1, 2, 2, 3)$ has a similar property in that

$$1^2 + 2^2 + 2^2 = 3^2.$$

Find some more quadruples with this property, and try to establish some patterns in your results.

COMMENTARY

National Curriculum

Familiarity with **square numbers** is necessary, which is level 5 (AT5). Forming and generalizing the sequences of solutions is level 6, but for computing and the algebraic notation to be used, level 7 is more appropriate (AT 5, 6).

Materials

Squared paper is useful for tabulating sequences.

GENERAL COMMENTS

Square numbers and their interconnections have been a source of fascination for thousands of years. Numerous other investigations could be framed in this context, for example 'find some numbers which can be the hypotenuse of an integer-sided right-angled triangle in more than one way'.

A good source for further ideas is Beiler[1].

SEARCHING FOR QUADRUPLES

Writing $x^2 + y^2 + z^2 = r^2$ in the form $x^2 + y^2 = r^2 - z^2$ gives a very productive strategy for the initial search, i.e. to look for the sum of two square numbers which is also the difference between two square numbers.

The square numbers are 1, 4, 9, 16, 25, 36, 49, 64, 100, . . .; since $1 + 4 = 9 - 4$ then a quadruple is found, this being the one in the statement of the problem. Noticing that $4 + 9 = 49 - 36$ gives another quadruple $(2, 3, 6, 7)$.

That all odd numbers from 3 upwards are the difference between two squares, $4 - 1$, $9 - 4$, $16 - 9$, . . . and likewise all multiples of 4 from 8 upwards, makes the search more systematic. More quadruples such as $(1, 4, 8, 9)$, $(3, 4, 12, 13)$, $(2, 5, 14, 15)$ and $(2, 4, 4, 6)$ are found, the last being double $(1, 2, 2, 3)$.

Another approach which has been employed is a fairly simple computer search using nested loops and a check condition.

It has already been noted that quadruples with all even numbers are possible by doubling any existing quadruple. However, none with all *odd* numbers have been found. How can this be explained?

SEQUENCING SOLUTIONS

Sorting the quadruples into sequences suggests patterns. For example the sequence

$(1, 2, 2, 3)$, $(2, 3, 6, 7)$, $(3, 4, 12, 13)$, . . .

can clearly be extended to $(4, 5, 20, 21)$, $(5, 6, 30, 31)$ and so on, which can be checked as valid quadruples. Pupils with some algebraic experience generalize this sequence in the form $(x, x + 1, x(x + 1), x(x + 1) + 1)$. Sixth-form students and beyond can check that this form fits the quadruple requirement.

Amongst other sequences which have been found and generalized are those starting

$(1, 4, 8, 9)$, $(2, 5, 14, 15)$, . . . and $(1, 2, 2, 3)$, $(1, 4, 8, 9)$, . . .

LINKS WITH CUBOIDS

Articles in mathematical journals have appeared from time to time about these quadruples, mainly in connection with the geometrical aspect. Extending Pythagoras' theorem into three dimensions shows that quadruples describe integer-sided cuboids in which the space diagonal is also an integer. An extra constraint can be put on the quadruple by requiring that all the face diagonals of the cuboid are also integers, but no such 'perfect' cuboid has yet been found, nor is it known whether it exists.

Some 'imperfect' ones exist such as $(3, 4, 12, 13)$ which those familiar with Pythagorean triples will immediately see that one face has a diagonal of length 5. Other such cuboids, for example $(5, 12, 84, 85)$, can fairly easily be found by applying a search method along lines already described to the first two numbers in known Pythagorean triples.

Cuboids which are nearer to perfection in having two face diagonals as integers (as well as the space diagonal) do exist but are extremely difficult to find. The smallest has a space diagonal of 697.

An heuristically conducted computer search should be within the scope of computer-oriented mathematical students.

GENERATING QUADRUPLES

A general method of generating all quadruples is given by

$$x = \frac{a^2 + b^2 - c^2}{c}, \ y = 2a, \ z = 2b, \ r = \frac{a^2 + b^2 + c^2}{c}$$

where a, b and c are integers and provided c is a factor of $a^2 + b^2$ and $c < \sqrt{a^2 + b^2}$. This is an adaptation of an earlier general solution,

$$x = a^2 + b^2 - c^2, \ y = 2ac, \ z = 2bc, \ r = a^2 + b^2 + c^2$$

in which, perhaps surprisingly, a, b and c need not necessarily be integers or even rational.

REFERENCES

1 Beiler, A.H. (1966) *Recreations in the Theory of Number*, Dover, chapters 14, 15.

35 DESIGN

STATEMENT OF PROBLEM

The simplest polyhedron is the tetrahedron or triangular-based pyramid
which has four triangular faces as
shown. Another type of pyramid also
has four triangular faces together
with a quadrilateral face, its base.

Design (and construct if you wish) a
polyhedron with four triangular faces
and *two* quadrilateral faces.

Specify and design polyhedra with other combinations of faces.

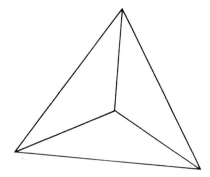

COMMENTARY

National Curriculum

Constructing 3-D shapes, such as cubes and pyramids, from **nets** is level 4 (AT10). The
investigation, though based on these skills, involves a development of them. This, and the
reasoning involved suggest level 5 or 6. Using 2-D representations of 3-D objects is level 6.

Materials

The usual construction materials are needed: firm card, instruments for constructing nets and
scoring folds, and a quick-setting glue.

Squared paper is useful for tabulating results.

GENERAL COMMENTS

Traditionally, this type of activity concentrates on regular and semi-regular polyhedra,
whereas the development of this study is towards irregular polyhedra and the conditions
which apply to them in terms of their number and type of faces, and numbers of edges and
vertices. It further encourages that important design skill of being able to visualize a three-
dimensional figure which has been drawn in two dimensions.

Initial ideas

Experience seems to suggest that a polyhedron must be visualized before its net can be constructed. This implies that to attempt to design the polyhedron specified in the statement of the problem by drawing a supposed net is fraught with difficulty. The best method of attack which has been found is to work with familiar polyhedra and truncate them in various ways; this creates an item bank of possible polyhedra which can be tabulated.

Truncating a tetrahedron

The required polyhedron has six faces and, since two faces meet at each edge, it has $10 \; [= \{(4 \times 3) + (2 \times 4)\} 12]$ edges. From the well-known Euler relation, $F - E + V = 2$, it follows that $V = 6$.

Truncating a tetrahedron can be done in three different ways, chopping off a vertex, or a side, or a case intermediate to these two. The following diagram shows the three cases with the specifications of the resulting polyhedra.

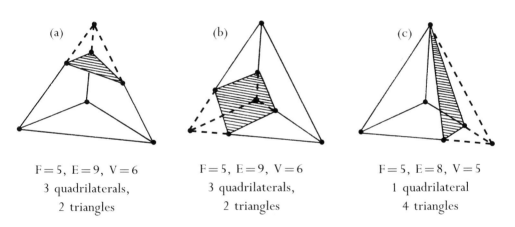

(a) F = 5, E = 9, V = 6
3 quadrilaterals,
2 triangles

(b) F = 5, E = 9, V = 6
3 quadrilaterals,
2 triangles

(c) F = 5, E = 8, V = 5
1 quadrilateral
4 triangles

Curiously, the first two new polyhedra are in principle equivalent, though this is not obvious at a cursory glance. Neither is it always realized that they are equivalent to the familiar triangular prism. The third polyhedron is the already familiar quadrilateral-based pyramid. The first truncations have not given rise to anything unfamiliar.

Truncating a different pyramid

Analogous truncations of a quadrilateral-based pyramid are shown in the next diagram, and the first truncation, (a), produces the required polyhedron.

Truncations (b) and (c) of this pyramid reveal a new aspect to the problem, in that two distinct polyhedra can have the same numbers of faces, edges and vertices.

(a)

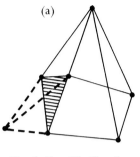

(b)

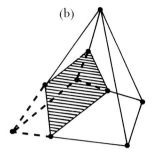

(c)

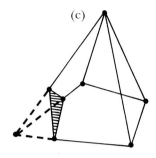

F=6, E=10, V=6

2 quadrilaterals,
4 triangles

F=6, E=11, V=7

4 quadrilaterals,
2 triangles

F=6, E=11, V=7

1 pentagon,
2 quadrilaterals,
3 triangles

FURTHER DESIGNS

As further examples are added to the list of possible polyhedra it may be remarked that some combinations of types of face do not occur. No polyhedron with, say, two quadrilateral and three triangular faces will be found. The number of edges is at least 6 and none will be found with just 7 edges. Some of these observations can be easily explained, others need some theory developed from Euler's relation and this is appropriate for older pupils only.

TOWARDS A THEORY

For plane-faced polyhedra

1. the minimum number of edges at a vertex is 3,
2. the minimum number of edges around a face is 3,
3. each edge joins 2 vertices, and
4. each edge is common to 2 faces.

Combining (1) and (3) shows that the number of edges is at least $3V/2$, and combining (2) and (4) shows that the number of edges is at least $3F/2$. Hence $2E \geqslant 3V$ and $2E \geqslant 3F$, so substituting in $F-E+V=2$ gives $E \geqslant 6$.

Further arguments can be developed to show that E can never equal 7 and, for example, that if the face with the smallest number of edges is a pentagon then there must be at least 12 such faces (the regular dodecahedron fits this theory).

CURVED FACES

The investigation could be extended to include curved faces, for which a simple example would be a slice of orange which has $F=3$, $E=3$ and $V=2$. Euler's relation still applies but not the above theory.

STATEMENT OF PROBLEM

A vector such as $\begin{pmatrix} 3 \\ 1 \end{pmatrix}$ describes a path on squared paper.

$\begin{pmatrix} 3 \\ 1 \end{pmatrix} \rightarrow \begin{pmatrix} 1 \\ 2 \end{pmatrix}$ means follow path $\begin{pmatrix} 3 \\ 1 \end{pmatrix}$ and then follow path $\begin{pmatrix} 1 \\ 2 \end{pmatrix}$.

The four vectors in the following arrangement will result in going round and round the quadrilateral drawn on squared paper.

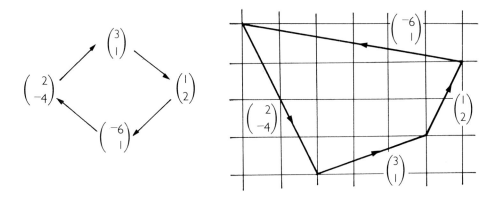

What happens if you arrange the vectors in a different order?
Experiment with other arrangements of vectors.

COMMENTARY

National Curriculum

Use of **vector** notation is level 8 (AT11); this however seems too high a level since level 4 refers to **coordinates** and level 6 to **bearings**. Use of negative elements in the vector implies level 5 as with coordinates.

Simple use of computer graphics could be employed, for example using the MOVE and DRAW commands in BASIC, which is level 4 (AT7).

Materials

Squared paper is vital throughout.

INITIAL APPROACHES

It will not immediately be realized that the sum of the four given vectors is the zero vector.

Pupils should be encouraged to write down *any* three vectors in a triangular arrangement as shown here. Since these do not add up to zero, going round and round their triangle will result in a continuous path of three parallel sets of vectors (strictly, equal free vectors), but not the expected triangle. This prompts the question of how to choose the vectors so that a triangle is formed.

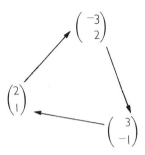

The 24 possible arrangements of the four given vectors give rise to just six distinct quadrilaterals, as each can be drawn in four different ways depending on which path is taken first. Two of the quadrilaterals are crossed which may cause some concern and need for redefinition of a quadrilateral. (See the investigation RANDOM DOTS (10).)

A very neat method which was devised for drawing all six of these quadrilaterals in one diagram is shown below, the vector $\begin{pmatrix} 3 \\ 1 \end{pmatrix}$ being drawn once only.

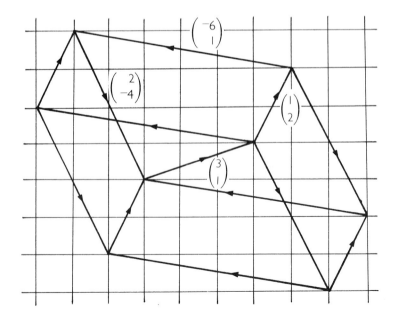

GOING ROUND BOTH WAYS

A good development is to make the arrows in the arrangement double-ended. This diagram shows three vectors **a**, **b** and **c** whose sum is the zero vector. To start, choose *a* and then at each successive move one of two can be chosen, since **b** or **c** can follow **a** and so on.

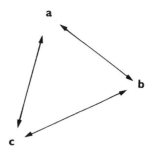

A coin can be used to make the random selection. The path can start at the centre of the paper and will eventually go off the side, at which point another path is started at the centre. Once a few paths have been drawn a tessellation of triangles starts to appear.

What pattern is produced if the arrows connecting the four vectors in the statement of the problem are double-ended?

OTHER TESSELLATIONS

Finding tessellations produced in this manner opens up the question of whether any tessellation could be described by a vector arrangement. The hexagon tessellation is straightforward and, with ingenuity, the semi-regular tessellations can be described. For example, the square and octagon (not regular) tessellation has been ingeniously described by the arrangement in this diagram.

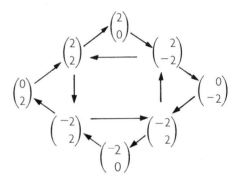

For older pupils and students, this particular method provides an intriguing challenge in graphics programming.

37 PIVOT POINTS

STATEMENT OF PROBLEM

The tower on the left below will fall onto its side (the dotted position) if it pivots about X. If you are in any doubt about this, check it with tracing paper or by cutting out. Use the point of a pencil as the pivot. The tower has turned through 90°.

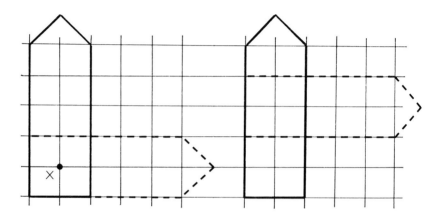

Can you say, about which point the tower on the right would have to pivot, in order to fall into the dotted position?

Try some other positions for the fallen tower and find the pivot points. Then try some upside down or other positions of the tower.

COMMENTARY

National Curriculum

This investigation is linked to the rotation of figures, in particular **centres of rotation**, and this is first mentioned at level 4 (AT11). However much of the development described here is by older pupils working at about level 8.

Materials

Squared and tracing paper are essential. Using a pencil as suggested above for the pivot point is preferred to the point of a pair of compasses.

INITIAL APPROACHES

Experimenting with special cases of figures based on squared paper enables results to be established fairly easily, and these can then be progressively generalized. Practical work with tracing paper as suggested in the statement of the problem is essential in order to develop some conjectures. The next two sections show just one way in which the exploration developed. Other ways can undoubtedly be found.

LINES OF PIVOT POINTS

The pivot point for the tower on the right in the statement of the problem should not present much of a problem. An obvious next stage would be to try the fallen tower in a position half-way between the two shown positions. Not surprisingly the new pivot point turns out to be half-way between the two known pivot points. Further positions could then be tried but keeping the base of the fallen tower along the same vertical line, namely the left-hand side of the upright tower. The pivot will always lie on a diagonal line. Why should this be and what determines the position of this line?

   The fallen tower could then be moved from the position on the left in a horizontal direction, into various other positions; the pivot point is now located on a different diagonal line. What determines the position of this diagonal line?

INCLINED POSITIONS

The line of development suggested in the previous section eventually results in the generalization shown in this diagram. Here the tower has fallen to an inclined position. Two rhombuses are formed, one by the longer sides of the rectangular part of the tower and the other by the shorter sides. It is argued that the pivot point lies on the two dotted diagonals of the rhombuses; it is on one diagonal to ensure that the pair of longer sides rotate correctly and on the other diagonal to ensure that the shorter sides rotate correctly.

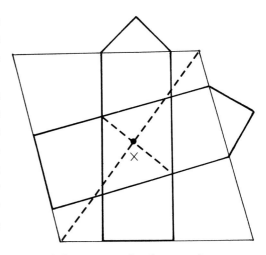

   Each of the two rhombuses has two diagonals and the reason why the two shown are selected to give the centre of rotation needs to be clarified.

STATEMENT OF PROBLEM

A rectangular table 1 m long and 0.5 m wide is being carried along a corridor 1 m wide in which there is a right-angle corner as shown. If the top of the table is kept horizontal can it be manoeuvred successfully round the bend?

Could the table be bigger and still be moved round the bend? A circular table of diameter 1 m will clearly work. Design some other shapes of table that will go round the bend.

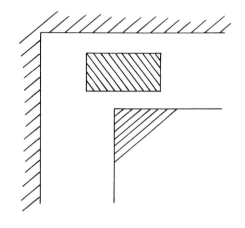

COMMENTARY

National Curriculum

Level 4 is the earliest for tackling this problem, and maximization arguments based on area considerations imply about level 6 (AT8), although the theoretical arguments using trigonometrical ideas imply level 8 (AT10) and upwards.

Materials

Squared paper is generally essential, and for practical work tracing paper and possibly card for cut-out plan models are useful.

GENERAL COMMENTS

This problem has received occasional attention in mathematical journals since 1966 at least, usually at an erudite level. It has become known as the 'sofa' problem for reasons which will become clear in due course.

For pupils it presents an intriguing two-dimensional real practical maximization problem which initially can be explored experimentally. Geometrical arguments can be used to explain maximum cases, and there is plenty of scope for advanced and higher level students to use **trigonometry** and **calculus**.

A SIMPLER PROBLEM

Start by considering the analogous problem of a ladder being carried round the corner whilst held in a horizontal position. To some it will appear intuitively clear that the critical position is the 45° one, with the ladder in contact with the inner corner and touching the two outer walls. Others may need to measure or calculate the maximum possible length of the ladder at other angles.

TURNING TABLES

A special case of rectangular table is often spotted, namely a square one of side equal to the width of the corridor which will slot neatly into the junction and so goes round the bend without turning! For any table longer than the width of the corridor then turning is bound to occur, and all such tables are generalizations of the ladder case. Once again the 45° position appears to be critical, since for a given length of table the corresponding maximum width is that which will just fit in the 45° position. The following diagram shows three tables of critical size.

To the table ABCD corresponds a ladder AB being carried round a junction of narrower corridors with the same outer walls and the inner corner being P.

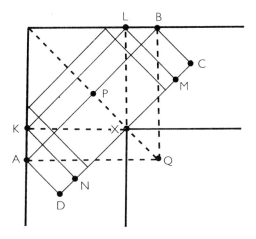

The table KLMN is special as two of its vertices touch the outer wall directly opposite the inner corner X. Experiment and theory both show that such a table will go round the bend by rotation about X with the middle point of one of its longer sides in contact with X.

On a similar principle, the table ABCD can be envisaged rotating about the point Q.

MAXIMIZING THE TABLE SIZE

Which of this set of tables has the maximum area? It is fairly easy to see that the area of KLMN is 1 m² either by dissection or otherwise. Since its proportions are 2 by 1 it can be argued that table ABCD has less area as more has been cut off KLMN along one of the longer sides than has been added on to the two shorter sides. A similar argument applies to the other table (can that one go round the bend?) and so KLMN is the maximum. Unfortunately, this is no bigger than the square table which was mentioned at the beginning of the previous section.

OTHER SHAPES OF TABLE

Allowing alternative shapes for the table gives great scope for ingenuity. A popular example is a semi-circular table of radius 1 m which rotates neatly on the inner corner of the junction. Having an area greater than 1 m² this is a distinct improvement on the rectangular case.

The erudite approaches mentioned above extend this case by including a rectangular section of table, with a semi-circular cut-out, between the two halves of a semi-circular table. Such a table (or sofa, from its shape) is shown in this diagram part-way round the bend. Note that the two 'arms' of the sofa stay in contact with the outer walls whilst the semi-circular cut-out stays in contact with the inner walls and corner. If the dimensions of the additional middle part are chosen to maximize its area then it is claimed that an overall maximum for the problem is achieved.

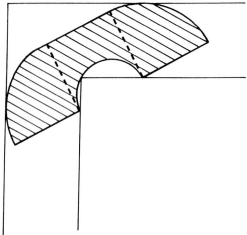

YET ANOTHER TABLE

An alternative approach suggested and explored by pupils is to link two circular tables of the type mentioned in the statement of the problem by a joining section to form one composite table. This seems to be a more obvious line of development than the sofa approach.

EXTENSIONS

Other variations which have been considered are:

- corridors of differing width, meeting at 90° – this generalization, from corridors of equal width, does not have the critical position for rectangular tables at 45°;
- equal width corridors meeting at different angles – here the semi-circular table still suffices, but for quite acute angles a sector shape with its point in the outer corner becomes a better result.

39 SQUARE SPIRAL

STATEMENT OF PROBLEM

On an A4 sheet of centimetre-squared paper draw a square of side 16 cm. Draw a line to halve it and shade the left half, as in the left-hand diagram below.

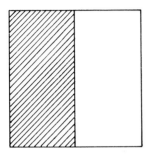 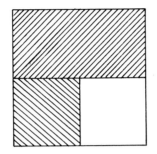 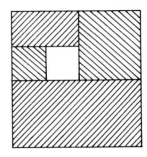

Turn the paper through 90° clockwise, draw a line to halve the unshaded part of the figure and shade the left half, as in the centre diagram. Repeat this process as many times as you can realistically; the right-hand diagram shows the state after two more repeats.

As the unshaded part gets smaller and smaller where does it eventually end up?

Create another spiral of shaded rectangles, by starting with a rectangle rather than a square.

Can you devise a spiral based on a triangle?

COMMENTARY

National Curriculum

This investigation is designed to explore the concept of a **limit**. In this respect it shares some common ground with the investigation FIGURE SEQUENCES (17). In the commentary on that investigation it was remarked that the limit concept is only implicit in the National Curriculum at the higher levels of ATs 5 and 7. Working at level 7 and above, older pupils and students should be able to use analytical and geometrical ideas to determine the precise position of the limit. Other ideas involved are **ratio** (level 6, AT3) and **enlargement** and **similarity** (levels 6 to 8, AT11).

Nevertheless the skills involved in creating a successful drawing are well within the capabilities of much younger pupils at level 4 (AT10) and it is intriguing to read their ideas about where the limiting position is.

Materials

Squared paper is essential, as is triangled paper for triangle sequences.

INITIAL APPROACHES

It can be helpful to mark the centre points of each shaded rectangle or square and join them up in order as in this diagram. (Note that the square is drawn in its initial orientation.) It is often remarked that alternate sections of this spiral are parallel. Why should this be so?

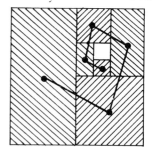

Perhaps more surprising is that these centre points lie on one of two lines, and this observation should be explained. It then seems quite reasonable to argue that the limiting position is where these two lines intersect.

FOCUSING ON A DIAGONAL

Another line of argument develops as follows. After two shadings, the unshaded square lies on a diagonal of the initial square, and after another two shadings the unshaded square is again on the same diagonal and so on. Therefore the limiting position must be on the diagonal. Precisely where on the diagonal?

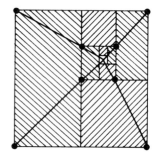

THE LIMIT AS A CENTRE OF ENLARGEMENT

If the corresponding vertices of the initial square and the unshaded squares after four and eight shadings are joined, the four straight lines concur at a centre of enlargement for the three squares as shown in the following diagram. Two of the four lines form the diagonal of the square referred to above.

This centre of enlargement must be the limiting position, and with the unshaded square, after four shadings, being a quarter of the original square, the position of the limit can be calculated.

OTHER SPIRALS

A rectangle start produces analogous figures to those already drawn.

A sequence based on a triangle could be started with an equilateral triangle of side 16 cm on triangled paper, the first line being drawn to cut off an 8 cm equilateral triangle, shading the remaining trapezium and so on.

40 FRACTION CHALLENGE

STATEMENT OF PROBLEM

- Amanda is trying to find a fraction between $\frac{2}{5}$ and $\frac{1}{2}$. First of all she changes them into tenths, $\frac{4}{10}$ and $\frac{5}{10}$, and then into twentieths, $\frac{8}{20}$ and $\frac{10}{20}$. She argues that $\frac{9}{20}$ is therefore between $\frac{2}{5}$ and $\frac{1}{2}$.

  Alison says she has a different method, in which she adds the tops and bottoms of the two fractions, $(2+1=3$ and $5+2=7)$ and claims that $\frac{3}{7}$ is in between $\frac{2}{5}$ and $\frac{1}{2}$.

  Is Alison right? If so, can you explain why?

- Amanda decides to use Alison's method to find a fraction which when multiplied by itself gives 2 (i.e. she seeks the square root of 2). She tries $\frac{3}{2}$ and $\frac{5}{4}$ but finds that the first fraction is too big as $\frac{3}{2} \times \frac{3}{2} = \frac{9}{4} = 2\frac{1}{4}$, and the second fraction is too small as $\frac{5}{4} \times \frac{5}{4} = \frac{25}{16} = 1\frac{9}{16}$. Using Alison's method she finds $\frac{3+5}{2+4} = \frac{8}{6} = \frac{4}{3}$.

  Is this fraction too big or too small? Can you find a better result?

COMMENTARY

National Curriculum

A good understanding of **fractions** is clearly a prerequisite for this investigation. The National Curriculum seems confused on this topic, with understanding equivalence of fractions put at level 6 in AT2, though level 4 of AT5 refers to exploring the same idea in the context of number pattern. Addition and multiplication of fractions have no explicit mention but are implicit from level 7 upwards in ATs 2, 3, 5 and 7. Distinguishing **rational** and **irrational** numbers is level 9 (AT2).

Materials

A calculator is useful to check the approximations.

GENERAL COMMENTS

The problem of approximating irrationals by rationals has a long history in mathematics. A recent and fascinating account is to be found in an article by Fletcher[1] which shows, *inter alia*, how Alison's method leads to the well-known approximation 5 miles $= 8$ kilometres and to the renowned and remarkable Chinese approximation of $\frac{355}{113}$ for π, which is correct to seven significant figures.

The method is linked, ingeniously, with continued fractions.

JUSTIFYING ALISON'S METHOD

Alison's method can be justified broadly by arguing that she is effectively finding the mean of the numerator and denominator separately. Proof by manipulating algebraic inequalities is clearly possible, or a neat geometrical illustration arises when fractions are graphed as ordered pairs, for example by representing $\frac{2}{3}$ by the point $(3, 2)$. In the diagram, the fractions $\frac{2}{3}$ and $\frac{1}{4}$ are represented by the vectors **OA** and **OB**. Alison's in-between fraction, $\frac{3}{7}$ is represented by **OC** which is the vector sum of **OA** and **OB**, and hence bound to lie between **OA** and **OB**. The fractions are the gradients of **OA**, **OB** and **OC**.

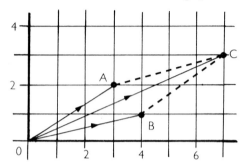

RATIONAL APPROXIMATIONS

Developing Amanda's search for a rational approximation to $\sqrt{2}$ is a rewarding exercise. The fraction $\frac{4}{3}$ is still too low but closer than $\frac{5}{4}$, so using Alison's method again, on $\frac{3}{2}$ and $\frac{4}{3}$ gives $\frac{7}{5}$. This is still too low, but draws closer.

If at each stage the closest high and low results so far obtained are used the sequence of fractions

$$\frac{3}{2} \quad \frac{5}{4} \quad \frac{4}{3} \quad \frac{7}{5} \quad \frac{10}{7}\ \frac{17}{12} \quad \frac{24}{17}\ \frac{41}{29} \quad \frac{58}{41}\ \frac{99}{70} \cdots$$

is produced. Some of these approximations are better than others, relative to the size of the numbers involved. Those in which the square of the numerator only differs by 1 from twice the square of the denominator are

$$\frac{3}{2} \quad \frac{7}{5} \quad \frac{17}{12} \quad \frac{41}{29} \quad \frac{99}{70} \cdots$$

and a rule by which each of these can be determined from the previous one is not difficult to spot.

Why does this rule create successively closer approximations to $\sqrt{2}$? It is interesting to apply the same rule to the fraction $\frac{5}{4}$ and see what happens.

FURTHER DEVELOPMENT

After the first four fractions, the sequence of high and low estimates for $\sqrt{2}$ in the previous section is shown in pairs, two high, two low, two high, and so on. Fletcher[1] points out that this regular pattern will persist.

Suppose a sequence of rational approximations for $\sqrt{3}$ is calculated using the same principle and starting with $\frac{2}{1}$ and $\frac{3}{2}$ as high and low estimates. Does a regular pattern of the high and low estimates still occur?

Other irrational numbers such as $\sqrt[3]{2}$ can be explored, and it is also possible to find approximations for rational numbers which have large numbers of figures in the numerator and denominator. For example $\frac{40}{23}$ is a good approximation for $\frac{98765}{56789}$, being in agreement on the first five significant figures.

REFERENCES

1 Fletcher, J.J. (1974) Approximating by vectors, *Mathematics Teaching*, **63, 64**.

INDEX